PHYSICAL SCIENCE I

Physical Science I

Rhys Lewis

B.Sc.Tech., C.Eng., M.I.E.E., A.M.C.S.T.,

*Head of Electronic and Radio Engineering Department,
Riversdale College of Technology,
Liverpool*

First published 1978 by
THE MACMILLAN PRESS LTD
London and Basingstoke
Associated companies in Delhi Dublin
Hong Kong Johannesburg Lagos Melbourne
New York Singapore and Tokyo

Printed and bound in Great Britain by
A. Wheaton & Co. Ltd., Exeter

British Library Cataloguing in Publication Data

Lewis, Rhys
    Physical Science I.—(Macmillan technician series).
    1. Science
    I. Title   II. Series
    500.2'02'46   Q160.2

    ISBN 0–333–19219–2

Contents

Preface

This book contains the theory necessary for the student to attain all the general and specific objectives in TEC Standard Unit U75/004 Physical Science I, except of course the small number involving the use of equipment which can only be attained within a laboratory or workshop situation.

The book is divided into twelve chapters, covering all the main topic areas in the Standard Unit. All chapters contain a number of worked examples, and assessment exercises are provided at the end of each chapter. Each exercise contains long-answer, short-answer and multiple-choice questions. At the beginning of each chapter, where relevant, the general and specific objectives are given, the letter and number code being as given in the TEC Standard Unit.

The over-all arrangement of the book is such that a logical progression is made by the student from the known to the unknown, prerequisites for any particular chapter being covered in preceding chapters.

Acknowledgements

A personal tribute

I am grateful to the following firms for permission to use drawings, photographs or other material and for their assistance in producing a content as up to date as possible

The Ever Ready Co. (Great Britain) Ltd, Whetstone, London
Chloride Industrial Batteries Ltd, Swinton, Manchester
G. Cussons Ltd, Manchester (Scientific Equipment Manufacturers).

I would also like to thank J. Keefe, for his assistance in obtaining some of the diagrams, and, as always, my wife, for her much appreciated help in typing the manuscript and for her encouragement and other assistance during a rather prolonged period of preparation of the book.

RHYS LEWIS

The theory contained in this book is a necessary part of the basic knowledge of all engineers at whatever level and in whichever discipline they operate. The book is therefore a fitting place to record my personal tribute and gratitude to the late Harold Ashton, who for over a quarter of a century at Openshaw Technical College inspired and helped colleagues, student teachers and most of all his students. He will be long remembered not only for his considerable ability and innumerable skills as an engineer and teacher but as a tremendous personality and as a friend.

RHYS LEWIS

1 Units of Measurement: The International System

Most of us begin to measure and use units of measurement long before we even meet the words 'measure' and 'unit'. Very small children soon learn what a penny is and, before long, a second, a minute and an hour. We go on to learn of length and mass and weight (more about these later) and all of us, whatever our job may be, need to know what these quantities are, and what units they are measured in, every time we make a purchase and indeed every time we obtain materials to make anything.

Engineering is concerned with making—making machines, buildings, bridges, making the electrical and electronic equipment for them, making a whole variety of things for everyday use. Engineers, probably more than anyone, need to measure and need to understand measurement, what it is and how it is done.

WHAT IS MEASUREMENT?

A dictionary definition of *measure* is 'the extent, dimensions, capacity, etc., of anything, especially as determined by a standard'. The same dictionary* defines *standard* as 'something established as a rule or basis of comparison in measuring'. Although these definitions—like many dictionary definitions—depend on each other to some extent, the general idea is fairly clear. Measurement is determining the size of a quantity and is carried out by comparison with a standard or defined size of the same quantity. The standard or defined size is, of course, called a *unit*.

* *Webster's New World Dictionary*

SO MANY QUANTITIES AND SO MANY UNITS

Engineering students can be forgiven for getting confused. When we first begin we meet what is apparently a limitless number of quantities with an equally limitless number of units, many of them with strange names. What is not apparent is that most of these quantities are related to each other and, in fact, belong to a *system*.

WHAT IS A SYSTEM OF UNITS?

A system is a 'regular order of connected parts' (the dictionary again) and a unit is 'any known determinate quantity by constant application of which any other similar quantity may be measured'. The word *determinate* comes from the verb 'to determine', which also means 'to define'.

A system of units, then, is 'a regular or logical order of defined quantities by constant application of which other similar quantities may be measured'. Although there are many possible systems of units, all systems are based on only a small number of basic units.

Basic Units

The basic units are those of quantities which are 'intuitively fundamental'. Intuition means 'knowing without the conscious use of reason'. In other words, the quantity must be one we know without having to think about it in terms of long-winded definitions (they come later!).

The basic units chosen for most systems of units include the basic quantities *mass*, *length* and *time*. These, of course, are the ones we mentioned earlier when we considered how early in life we begin to measure, and, although there is often a confusion between mass and weight, these quantities are certainly intuitively fundamental. If you feel unsure about that, think of some other quantities, like speed, for example; speed is *not* intuitively fundamental—it connects distance (length) and time, which are two of the quantities which *are* regarded as intuitively fundamental. Mass, length and time are not the only basic quantities—we shall examine others later.

WHICH SYSTEM?

Continental countries have used metres and kilograms for many years (the M.K.S. System); in the United Kingdom and in the United States and Canada we have used feet and pounds (the F.P.S. or Imperial Gravitational System). The unit of time, the second, is, of course, international. Apart from the fact that the use of different units on a world-wide scale leads to some confusion, particularly now that engineers are more involved in international projects, many of the systems derived from the basic units are scientifically suspect. A unit is only of use if it can be accurately defined and used regardless of place, and many units—for example, the pound-force or pound-weight and the kilogram-force or kilopond—are dependent on variables such as the force of gravity, which changes from place to place. A system dependent on gravity for its definitions is called a *gravitational* system, whereas a system not so dependent is called an *absolute* system. Clearly, an absolute system is preferable to a gravitational system of units.

We said earlier that a system of units is built up from a few basic units. We have already met an example in speed, which connects distance or length with time, so that the unit of speed will contain length units and time units. If we define the unit of speed as *one* distance unit per *one* time unit (metre per second or foot per second, etc.) then the derived unit, the unit of speed, is called a *coherent* unit. Many derived units are not coherent and we have to remember values of connecting constants. From the point of view of simplicity, it would be a good idea to choose a system of units containing as many coherent units as possible.

To measure quantities which are larger or smaller than the unit size we use *multiples* and *sub-multiples*—inch, yard, mile, etc., for the Imperial System, centimetre, kilometre, etc., for the International System. Imperial System multiples and sub-multiples are often related by apparently unconnected constants (1 foot = 12 inches, 1 yard = 3 feet, etc.) but the International System is a decimal system and its multiples and sub-multiples are related by powers of ten (1 kilometre = 1000 metres, 1 metre = 100 centimetres, etc.). The names of these are given later.

The International System is absolute, coherent and has easy to remember multiples and sub-multiples. The Imperial System, until fairly recently in common use, is not absolute or coherent and its multiples and sub-multiples are not as easy to remember.

The International System of Units

The International System of Units is a development of a decimal system proposed by a Frenchman, Simon Stevin, over 300 years ago. A number of years elapsed before the system he proposed developed into anything like the one we now have and it was 1795 when the metric system based on the centimetre, gram and second was decreed by law in France. The Metre convention recognising the metric system was eventually signed by seventeen countries, all of which, except the United Kingdom and the United States, went on to adopt the system for everyday use. Later, the basic mass unit was changed to the kilogram (1000 grams) and the basic length unit to the metre (100 centimetres) and the system became known as the M.K.S. System. For a number of years, however, the system in general use in Europe was the M.K.S. Gravitational System with some of its units dependent on gravity, and it was relatively recently (1954) that an M.K.S. Absolute System was adopted by the Conférence Générale des Poids et Mesures (an international body concerned with units and unit systems). The title 'Système International d'Unités', abbreviated to SI, was adopted in 1960.

At the time of writing, the International System is still not truly international. Some European countries still use the M.K.S. Gravitational System although the move towards SI is very advanced (France declared all other systems illegal in 1962). The United Kingdom is in the process of 'metrication', which is an effective adoption of SI, and the United States and Canada are giving the matter a great deal of thought.

DEFINING THE BASIC UNITS

We said earlier that a basic unit size must not depend on location and this is why an absolute system of units is preferable to a gravitational system. It is also a good idea to define a unit size so that it can be measured fairly easily throughout the world; the possibility of inaccuracy of measurement would be much reduced by this means.

Originally the metre was defined as one ten-millionth part of the distance from pole to equator along the meridian passing through Paris. This is as good a definition as anything since presumably Paris and the pole do not move relative to one another. The only problem is—what is the distance? Over the years, as methods of measurement changed, so apparently did the distance between the pole and Paris! A similar problem occurred with the definition of the time unit, the second, taken originally as 1/86400th part of the day. How long is a day? It does not really help to say 'a mean solar day' (as they did later) since, again, a further definition is needed. They did not have the same trouble with the kilogram since this was defined as a certain mass, the standard of which (called an 'etalon') is kept at Sevres in France. The only problem with doing this is, if any query arises over mass sizes, the only *ultimate* solution is to go to France.

Nowadays we rely on the atom for two of the basic definitions. As you probably know from schoolwork, all substances are made up of one or more basic *elements*, each element consisting of even smaller parts called *atoms*. The atom is the smallest part of an element to retain the element's individuality—that is, although any atom may be further broken up, if it is it will not be an atom of the particular element. The atom itself gives out electromagnetic radiations (similar to light waves) under certain conditions and it is these radiations from a particular atom which are now used to define both the metre and the second. The kilogram unit unfortunately is still dependent on a standard size as before.

The unit definitions are given at the end of this chapter. It is not necessary to commit them to memory as long as the principles behind them are understood—that is, that any definitions should involve reproducible measurement with an accuracy independent of place.

The Other Basic Units

So far we have mentioned only three basic units. The SI System, in fact, is based on six, five of which are truly independent. The remaining three are units of temperature, electric current and luminous intensity (a quantity connected with light). The light unit actually depends on some of the other basic units but is nevertheless taken as one of the basic six.

Discussion of these other units is contained in the chapters on heat and electricity. Definitions of all six units are given at the end of this chapter; all other SI units are derived from these six.

MECHANICAL QUANTITIES

The units of all mechanical quantities are derived from the units of mass, length and time. The important mechanical quantities arranged in order of derivation are

speed
acceleration
force
work or energy
power

All these quantities and their units are discussed in detail in the chapters devoted to them; this chapter serves only as an introduction.

Speed is the rate of change of distance (length) with time. The unit is the metre per second, abbreviated to m/s.

Acceleration is the rate of change of speed with time. The unit is the metre per second per second, abbreviated to m/s^2.

Force (defined on p. 22) is measured in units given by multiplying mass by acceleration. The unit is called the *newton*, abbreviated to N, where

one newton = one kilogram-metre/second^2

Energy is measured in units given by multiplying together force and distance. The unit is the *joule*, abbreviated to J, where

one joule = one newton-metre

Power is the rate of using energy and is measured in joules per second. One joule per second is called one *watt*, abbreviated to W, where

one watt = one joule/second

The connection between mechanical units is shown in figure 1.1.

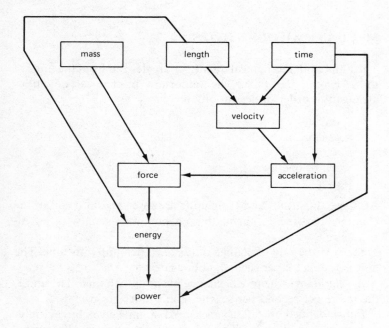

Figure 1.1

MULTIPLES AND SUB-MULTIPLES

As we said earlier, large or small quantities may be measured in multiples or sub-multiples of the unit used for the particular quantity. These multiples and sub-multiples have standard names and symbols written *before* the unit symbol. The multiples and sub-multiples in SI are all powers of ten, that is, 10, 100, 1000, etc., or 1/10, 1/100, 1/1000, etc.

In table 1.1, the number of noughts in the multiplier is written in

small figures to the right of 10 and slightly above the line. No sign before the power means a *positive* power, thus 10^2 means 100, 10^3 means 1000, and so on. A minus sign before the power means a *negative* power, thus 10^{-2} means 1/100, 10^{-3} means 1/1000, and so on.

Table 1.1

Prefix	Abbreviation	Multiplier
tera	T	10^{12}
giga	G	10^9
mega	M	10^6
kilo	k	10^3
hecto	h	10^2
deca	da	10^1
deci	d	10^{-1}
centi	c	10^{-2}
milli	m	10^{-3}
micro	μ	10^{-6}
nano	n	10^{-9}
pico	p	10^{-12}
femto	f	10^{-15}
atto	a	10^{-18}

Note Only one prefix should be used with any unit, thus pico rather than micromicro.

THE BASIC SI UNITS

The metre is the length equal to 1 650 763.73 wavelengths in vacuum of the radiation corresponding to the transition between the levels $2P_{10}$ and $5d_5$ of the atom of krypton-86.

The kilogram is the unit of mass; it is equal to the mass of the international prototype of the kilogram.

The second is the duration of 9 192 631 770 periods of the radiation corresponding to the transition between the two hyperfine levels of the ground state of the atom of caesium-133.

The ampere is that constant current which, if maintained in two straight parallel conductors of infinite length, of negligible circular cross-section, and placed one metre apart in vacuum, would produce between these conductors a force equal to 2×10^{-7} newtons per metre of length.

The kelvin (unit of thermodynamic temperature) is the fraction 1/273.26 of the thermodynamic temperature of the triple point of water.

The candela is the luminous intensity, in the perpendicular direction, of a surface of 1/600 000 square metres of a black body at the freezing point of platinum under a pressure of 101 325 newtons per square metre.

Supplementary units to be used with these are: the radian for angles (and steradian for solid angles) and the mole for 'amount of substance'.

2 Motion

OBJECTIVES

*All the objectives should be understood to be prefixed by the words
'The expected learning outcome is that the student . . .'*

E14 Solves problems involving uniform motion and acceleration.

14.1 Defines speed.

14.2 Calculates speeds from given time and distance data.

14.3 Plots distance–time graphs.

14.4 Calculates the slope of such graphs and interprets the
slope as speed.

14.5 Calculates average speeds from given numerical and
graphical data.

14.6 States the difference between speed and velocity.

14.7 Defines acceleration.

14.8 Plots speed–time graphs for motion in a straight line.

14.9 Calculates the slope of such graphs and interprets the
slope as acceleration.

14.10 Describes 'free fall' as being constant acceleration.

14.11 Solves problems using the equation s = average velocity
$\times$ time.

14.12 Solves problems using the equation $v = u + at$, including
motion under gravity.

14.13 States that the unit of force is the newton.

14.14 States that acceleration is the result of a net force being
applied.

Any body or part of a body which changes its position is said to be in *motion*. Motion means movement from one place to another. Motion may be described in terms of *speed* and *velocity* (which are not the same thing) and changing motion in terms of *acceleration*. A *change* of motion is caused by the application of a *force*.

SPEED

The speed of a moving body or part of a body is the *rate of change* with time of the distance of the body or part from some fixed point. The idea of 'rate of change' is very important in engineering and it will help if we look more closely at what it means.

Suppose we have the information that a car leaving Manchester at 9.00 a.m. arrives somewhere in London, a distance of 300 kilometres away, at 2.00 p.m. the same day. What is the car's speed? All we know is that the car travels 300 km in 5 hours. The change of distance of the car from Manchester is 300 km per 5 h. What then is the rate of change of this distance with time? The answer, briefly, is that we do not know and cannot say—it all depends on how the car travels to London. Figures 2.1, 2.2 and 2.3 show three different ways in which the car could make the journey. All the figures are in the form of graphs plotting distance from Manchester (km) vertically and time taken from start (h) horizontally. Any point on the graph corresponding to any particular time gives the distance covered by that time; for example, in figure 2.3 the point marked

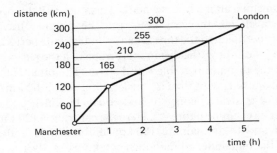

Figure 2.2

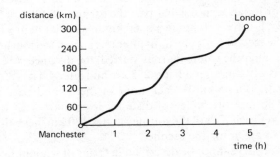

Figure 2.3

'London' read down to the time axis gives 'time = 5 h', and read along to the distance axis gives 'distance = 300 km'. The first graph is a straight line joining 'Manchester' and 'London'. The second graph consists of two straight lines, the first one joining 'Manchester' to the point at which distance is 120 km and time 1 h, the second one joining this point to 'London'. The third graph is very interesting but indescribable—it is just a wavy line joining the beginning and end of the journey! Let us look at these figures more closely.

In figure 2.1 we see that after 1 hour the distance is 60 km, after 2 hours 120 km, after 3 hours 180 km and after 4 hours 240 km, the

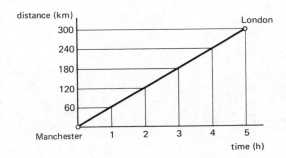

Figure 2.1

journey finishing after 5 hours when the distance is 300 km. Clearly this car travels 60 km every hour and the distance travelled is *directly proportional* to the time taken—that is, if the period of time considered is doubled, the distance is doubled, if the period of time is trebled the distance is trebled, and so on. The rate of change of distance from start with time is constant and equals 60 km/h, thus the *speed* of this car is constant and equals 60 km/h.

Now let us examine figure 2.2. Here the car travels 120 km in the first hour and the remaining 180 km in the next 4 hours, each 45 km of the remainder of the journey taking exactly 1 hour. The distance–time graph of part of the journey is a straight line and, as with the first case, over the period of each part of the journey, the distance between the car and its starting point is directly proportional to the time taken from the start of each part of the journey. For instance, in the first part the car travels 120 km in 1 h and, clearly, it travels 30 km in $\frac{1}{4}$ h, 60 km in $\frac{1}{2}$ h, 90 km in $\frac{3}{4}$ h, and so on. In the second part, one hour after *this part* of the journey has started (*not* after the whole journey started) the distance is 165 km, after 2 h it is 210 km, after 3 h it is 255 km and after 4 h it is 300 km, so the car travels 45 km in each hour of the second part of the journey. The rate of change of distance from start with time is constant over each part of the journey and is 120 km/h for the first hour and 45 km/h for the second.

Since figure 2.3 cannot be described in terms of straight lines we cannot say that the distance travelled changes at a constant rate; the rate of change is itself changing—the speed of the car is changing as the car is travelling. This case is in fact the closest to what normally happens on such a journey. Even on motorways where stopping is normally unnecessary, the speed of a car changes as the driver avoids getting too close to other traffic.

So we see that to obtain the speed of a moving body it is not sufficient merely to divide distance travelled by time taken. The figure obtained by doing that is called the *average speed*, which is the constant speed at which the body would have to travel to cover the given distance in the given time. Actual speed may vary from minute to minute, or second to second and the only true picture in such a case is that given by a graph of the form of figure 2.3.

So, given a distance–time graph how can we determine speed? Consider a body moving at a constant speed of (a) 10 km/h, (b)

20 km/h and (c) 5 km/h. The graphs of each of these cases are straight lines as shown in figure 2.4. We see that in each case over a time period of 1 h the line rises by an amount corresponding to the distance covered in 1 h. Dividing the value shown vertically by the value shown horizontally gives the speed of the body with the distance–time graph shown. Dividing the vertical by the horizontal in this manner gives the *slope* or *gradient* of the line. Thus the slope or gradient of the distance–time curve at any point gives the speed at that point. This applies *whether or not the speed is constant*. In figure 2.4d a part of a *non-linear* distance–time curve is shown. The slope at point P is the slope of the straight line shown and is equal to AB divided by BC. The speed of the body at the point in time corresponding to point P is AB/BC. If the distance–time graphs are linear the determination of speed is easy, if curved the determination of speed is not difficult but the accuracy may not be so good because of the slight differences in position that the straight line shown as AB in the figure can occupy (depending on who draws AB). AB is in fact the *tangent* to the curve at point P and there is only one correct position for it. If the distance–time

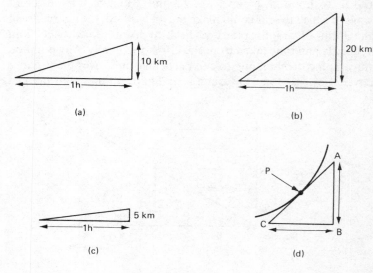

Figure 2.4

equation is known then the mathematics of rate of change (calculus) can be applied. We shall not, however, be doing that in this chapter.

Before we consider velocity, let us have a look at two problems involving speed.

Example 2.1

A car travels on a journey of distance 500 km in three stages, each of which is travelled at constant speed. The first stage takes 1 hour at a constant speed of 40 km/h, the last stage takes 3 hours and is of length 300 km. If the total journey takes 6 hours calculate (a) the speed during each of the second and third stages, (b) the average speed during stages one and two, during stages two and three and for the whole journey.

Solution In many problems of this nature it is best to gather together the known details and tabulate them.

	Distance	*Time*	*Speed*
Stage one	?	1 h	40 km/h
Stage two	?	?	?
Stage three	300 km	3 h	?
Total	500 km	6 h	?

(a) To calculate the speed during each of the second and third stages we need to know the distance covered and the time taken (and the fact that the speed is constant if we are to divide distance by time).

In stage two we know neither directly but we are able to determine both because of the information given for the whole journey and for stage one of the journey. The distance covered in stage one is 40 km since the car travels at 40 km/h for 1 h. The total journey length is 500 km (given) and the length of stages one and

three is 340 km (300 km given, 40 km calculated). Therefore the length of the second stage is $500 - 340 = 160$ km. The time for stage one is 1 h, for stage three is 3 h and for the whole journey is 6 h; thus the time for stage two is $6 - 3 - 1 = 2$ h. The speed for stage two (since it is constant) is therefore 160 km ÷ 2 h $= 80$ km/h.

The speed for stage three is 300 km ÷ 3 h $= 100$ km/h.

(b) Distance covered during stages one and two $= 40 + 160$

$$= 200 \text{ km}$$

Time taken $= 3$ h

Average speed $= \dfrac{200}{3}$

$$= 66.67 \text{ km/h}$$

Distance covered during stages two and three $= 300 + 160$

$$= 460 \text{ km}$$

Time taken $= 5$ h

Average speed $= \dfrac{460}{5}$

$$= 92 \text{ km/h}$$

Distance covered during whole journey $= 500$ km
Time taken $= 6$ h

Average speed $= \dfrac{500}{6}$

$$= 83.33 \text{ km/h}$$

Example 2.2

A car travels from town A to town B at an average speed of 48 km/h, from town B to town C at an average speed of 64 km/h and from C back to town A at an average speed of 50 km/h. The distances between towns A and B, between towns B and C and between towns C and A are equal. Calculate the average speed for the whole journey.

Solution To calculate the average speed for the whole journey we need to know the total distance covered and the time taken. We know the total distance is equal to three times the distance between any two of the towns (since the towns are equidistant) and we can find the time taken for each stage of the journey in terms of the distance covered during each stage of the journey. We do not know what this distance is, but suppose we let the distance between any two towns be x km. Then

$$\text{time taken between towns A and B} = \frac{x}{48}\,\text{h}$$

(since the speed between A and B is constant at 48 km/h) and

$$\text{time taken between towns B and C} = \frac{x}{64}\,\text{h}$$

(constant speed between B and C is 64 km/h) and

$$\text{time taken between towns C and A} = \frac{x}{50}\,\text{h}$$

(constant speed between C and A is 50 km/h). Thus

$$\text{total time} = \left(\frac{x}{48} + \frac{x}{64} + \frac{x}{50}\right)\text{h}$$
$$= 0.021x + 0.016x + 0.02x$$
$$= 0.057x$$
$$\text{total distance} = 3x$$

$$\text{Average speed} = \frac{\text{total distance}}{\text{total time}}$$
$$= \frac{3x}{0.057x}$$
$$= 53.14\,\text{km/h}$$

Note that the over-all average speed is *not* the sum of the individual speeds divided by 3 (which would give 54 km/h). The reason for this is that the time taken for each part of the journey is not the same.

Problems involving calculation of speeds and distances are included later in the chapter.

VELOCITY

Speed and velocity are terms which are often used to mean the same thing in everyday language. Strictly speaking they are not the same thing and, since engineers must be precise in their communication to avoid misunderstanding, it is essential to understand the difference between the two.

Velocity is the rate of change of distance during the time when the distance is measured *in a particular direction*. The units of velocity are those of speed but a direction must also be given if the name 'velocity' is used to describe the rate of change. Quantities that have magnitude (size) *and* direction are called *vector* quantities, those having only magnitude are called *scalar* quantities. Speed is a scalar quantity, velocity is a vector quantity.

To show the direction of action of a vector quantity in diagrams, lines are drawn in the appropriate direction and the diagrams are then called vector diagrams. To obtain a better understanding of velocity (and vectors in general) consider an aeroplane travelling due north at 800 km/h. If a wind rises blowing due east at 80 km/h we have a situation involving two vector quantities. In diagram form we could show this as in figure 2.5, showing 'north' as an upward direction and 'east' at an angle of 90° to the right. This diagram is drawn to *scale*, which means that the length of each line on the paper is proportional to the magnitude of the vector quantity the line represents. The scale is actually 1 mm = 20 km/h which means the aeroplane velocity (without the wind) is represented by a line 40 mm long ($40 \times 20 = 800$ km/h) and the line representing the wind velocity is 4 mm long ($4 \times 20 = 80$ km/h). A useful result of drawing scale diagrams is that we can obtain an idea at a glance of the relative size of vectors, and, in this case, the effect the easterly wind will have on the aeroplane. What effect is this? Well, clearly, without the wind the aeroplane continues in a northerly direction at 800 km/h but the easterly wind is going to

Figure 2.5

blow the aeroplane off course. The next question is by how much? To understand this we must learn what is meant by the term 'resultant velocity'.

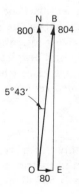

Figure 2.6

The resultant velocity of the aeroplane is the actual velocity (speed *and* direction) of the aeroplane when the engines are propelling it northwards at 800 km/h *and* the wind is blowing

eastwards at 80 km/h. The resultant velocity may be obtained by completing the parallelogram ONBE in figure 2.6 and joining OB. OB is then the resultant velocity, its length being proportional to the speed and angle NOB being its direction. Scale drawing gives the speed as 804 km/h in a direction N5°43′E. This method of finding resultants may be applied to any vectors and is called, appropriately enough, 'completing the parallelogram'.

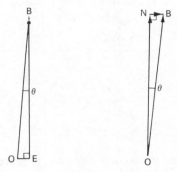

Figure 2.7

The same result may be obtained by drawing a vector triangle as shown in figure 2.7, in which the vectors are taken one by one (in any order) and the line representing the resultant is produced by joining the beginning of the first vector (point O) to the end of the second vector (point B). OB is the resultant at angle θ. The right-hand figure takes the aeroplane velocity first (ON), the wind velocity second (NB) and the resultant is OB at angle θ as before. Notice how a direction may be expressed by writing N5°43′E, which means north 5 degrees 43 minutes east, that is, the direction is at an angle 5°43′, the vector sloping away eastwards from the line representing the direction north.

In this example the actual velocity may be considered to have *component velocities* of 800 km/h due north and 80 km/h due east. The next two examples show, in the first one, how a resultant velocity is obtained and, in the second one, how component velocities may be calculated.

Example 2.3

A swimmer attempts to cross a river by swimming at right-angles to the bank. His normal swimming speed in still water is 0.3 m/s. If the river is flowing at 0.4 m/s find his resultant velocity.

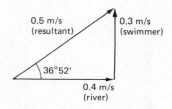

Figure 2.8

Solution   The vector diagram is shown in figure 2.8. By drawing, the resultant velocity is 0.5 m/s at an angle of 36° 52′ to the bank.

Example 2.4

If the swimmer in example 2.3 has a resultant velocity of 1.3 m/s at an angle of 22° 37′, find the velocity of the water and of the swimmer in still water.

Figure 2.9

Solution   The vector diagram is shown in figure 2.9. A line OS is drawn at an angle of 22° 37′ to a reference direction OA. Line SA is then drawn at right-angles to OA to meet line OA at A. Measure OA and AS and from the scale used determine the river velocity represented by OA and the swimmer velocity represented by AS. Scale drawing gives the river velocity as 1.2 m/s and the swimmer velocity as 0.5 m/s.

Further problems involving speed and velocity are included later in the chapter.

ACCELERATION

Acceleration is the rate of change of speed or velocity with time. As before, the rate of change may itself be changing from second to second and if the change in speed or velocity over a given period of time is divided by that time, the result is the *average* acceleration over that period. It is equal to the actual acceleration at any time only if the actual acceleration is constant over the period of observation. The units of acceleration are speed/time, for example, metres per second per second, written m/s². If acceleration concerns a change in velocity rather than speed then a direction is given and the acceleration will itself be a vector quantity; if only changing speed is considered then the acceleration has no given direction and is not a vector quantity. Acceleration may be determined from a velocity–time graph as shown in example 2.5. If the acceleration has a negative value (the moving body is slowing down) it is then usually called *retardation*.

Example 2.5

A car accelerates uniformly from zero to 50 km/h in 10 s. Determine the acceleration in metres per second per second.
Solution   The acceleration is 50 km/h per 10 s, which is

5 km/h/s

or

$$\frac{5000}{3600} \text{ m/s}^2$$

thus the answer is 1.38 m/s².

Example 2.6

A car moving at 53 km/h changes its speed to 77 km/h in 2 minutes. Determine the average acceleration over the period.

Solution   Change in speed = 77 − 53

$$= 24 \text{ km/h}$$
$$= \frac{24\,000}{3600} \text{ m/s}$$
$$= 6.67 \text{ m/s}$$

Average acceleration over 120 seconds $= \dfrac{6.67}{120}$

$$= 0.055 \text{ m/s}^2$$

Example 2.7

Calculate the constant value of retardation in m/s² if a train moving at 156 km/h is brought to rest in 2 minutes 33 seconds.

Solution

$$\text{speed} = 156 \text{ km/h}$$
$$= \frac{156\,000}{3600} \text{ m/s}$$
$$= 43.33 \text{ m/s}$$
$$\text{retardation} = \frac{43.33}{153} \text{ m/s}^2$$
$$= 0.283 \text{ m/s}^2$$

Example 2.8

A train travels with a uniform acceleration of 0.3 m/s² for 15 s and is then retarded at 0.6 m/s² for 22 s, after which it is at rest. Find the initial speed of the train and the greatest speed at which it travels during the motion.

Solution   This problem must be worked from the finish of the journey. If the train is retarded at 0.6 m/s² for 22 s and is then at rest, the speed at the beginning of the 22 s period is 0 × 22 m/s = 13.2 m/s.

Prior to this the train had been accelerated at 0.3 m/s² for 15 s, that is, the speed had changed by 0.3 × 15 m/s, that is, 4.5 m/s.

Thus the initial speed of the train is 13.2 − 4.5 m/s = 8.7 m/s and the greatest speed (immediately prior to the retardation period) is 13.2 m/s.

Example 2.9

Figure 2.10 shows a speed–time graph for a moving vehicle. Determine the least value, the greatest value and the mean value of the acceleration of the vehicle over the whole period.

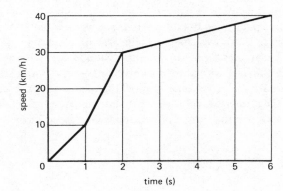

Figure 2.10

Solution   As can be seen, the graph is made up of three distinct sections, each a straight line. Each line shows speed changing at a *constant* rate, that is, the acceleration (the slope of the graph) is constant.

In the first second the speed change is 10 km/h. In the next second the speed change is 20 km/h (30 − 10) and in the next 4 seconds the speed change is 10 km/h.

The accelerations are 10 km/h/s in the first second, 20 km/h/s in the next second and 10/4 km/h/s, which is 2.5 km/h/s over the next 4 seconds.

The minimum acceleration is therefore 2.5 km/h/s and the maximum acceleration is 20 km/h/s. The minimum acceleration occurs over the least steep part of the graph and the maximum acceleration over the most steep part of the graph.

Over the whole period of 6 seconds the speed changes by 40 km/h. The average acceleration is therefore 40/6 km/h/s, that is, 6.67 km/h/s.

Area Under a Velocity–Time Graph

Consider a body moving with constant velocity u metres/second for a time t seconds. The distance covered during this period will be velocity × time, that is, ut metres. If a graph plotting velocity against time is drawn as in figure 2.11a it consists of a straight line parallel to the time axis as shown. We see that the area *under* the graph, shown shaded, is the area of a rectangle of sides u and t. The area is equal to ut, the distance covered in metres. Although this is a special case of a velocity–time graph the same applies to any velocity–time graph : the area under the graph is equal to the numerical value of the distance covered, provided that the time axis has the same units as those contained in the velocity

unit, that is, velocity in m/s and time in s yields distance in m; velocity in km/h and time in h yields distance in km, and so on. This fact relating distance and area is used in the next section concerning a body moving with constant acceleration.

Equations of Motion of a Body Moving with Constant Acceleration

Figure 2.11b shows the velocity–time graph of a body moving with constant acceleration. It is a straight-line graph because the gradient—which is equal to the acceleration—is constant. The body starts with velocity u and is accelerated with acceleration a for t seconds, after which the velocity is v. The change in velocity (acceleration × time) is at as shown on the graph.

From the graph

$$v = u + at$$

Now the area under a velocity–time graph is equal to the distance covered during the period covered by the graph. Let the distance be represented by s. Then

area under graph = area of rectangle (sides u and t)
            + area of triangle (sides t and at)
        $= ut + \frac{1}{2}at \times t$

thus

$$s = ut + \tfrac{1}{2}at^2$$

area under graph = area of trapezium with parallel
            sides u and v separated by
            side of length t
        $= \frac{1}{2}$ (sum of parallel sides) × distance
            between them
        $= \frac{1}{2}(u+v)t$

thus

$$s = \tfrac{1}{2}(u+v)t$$

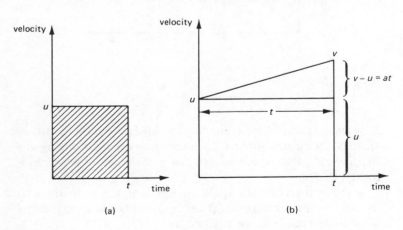

Figure 2.11

now

$$a = \frac{v - u}{t}$$

Multiply the left-hand side of the preceding equation by a and the right-hand side of the equation by $(v - u)/t$ so that the equation remains unchanged

$$as = \tfrac{1}{2}(v - u)(v + u)t/t$$
$$= \tfrac{1}{2}(v^2 - u^2)$$

thus

$$2as = (v^2 - u^2)$$

or

$$v^2 - u^2 = 2as$$
$$v^2 = u^2 + 2as$$

Summarising : if initial velocity $= u$, final velocity $= v$, acceleration $= a$, distance $= s$ and time $= t$ then

$$v = u + at$$
$$s = ut + \tfrac{1}{2}at^2$$
$$s = \frac{u + v}{2}t$$
$$v^2 = u^2 + 2as$$

These equations are most useful; it is a good idea to commit them to memory.

Example 2.10

A car is accelerated at 0.4 m/s^2 for 10 s from rest. Find the final velocity and the distance covered by the car.
Solution   Final velocity $=$ initial velocity $+$ acceleration $\times$ time

$$= 0 + 0.4 \times 10$$
$$= 4 \text{ m/s}$$
$$\text{distance} = [(0 + 4)/2] \times 10$$
$$= 20 \text{ m}$$

Example 2.11

A car travels a distance of 4.3 km, during which time it is uniformly accelerated from a speed of 25 km/h to 120 km/h. Find the acceleration.
Solution   Using $v^2 = u^2 + 2as$, where the symbols are as indicated above

$$v = \frac{120\,000}{3600}$$
$$= 33.33 \text{ m/s}$$
$$u = \frac{25\,000}{3600}$$
$$= 6.94 \text{ m/s}$$

Hence

$$33.33^2 = 6.94^2 + 2a \times 4300$$

$$a = \frac{33.33^2 - 6.94^2}{2 \times 4300}$$
$$= 0.1236 \text{ m/s}^2$$

Example 2.12

A vehicle is accelerated uniformly from rest to a speed of 15 m/s over a period of 24.3 s. Determine the distance covered by the vehicle during this period.
Solution

$$\text{Distance} = \tfrac{1}{2}(\text{initial speed} + \text{final speed}) \times \text{time}$$
$$= \tfrac{1}{2} \times 15 \times 24.3$$
$$= 182.25 \text{ m}$$

The equations of motion obtained above for a body being uniformly accelerated are also true for a body undergoing uniform angular acceleration, provided that for initial and final velocities we substitute initial and final angular velocities, for distance we substitute total angle turned through and, of course, for acceleration we substitute angular acceleration.

NEWTON'S LAWS OF MOTION

Most problems associated with everyday engineering may be solved by the application of Newton's laws of motion. Although we now appreciate that Newton's findings concerning motion and moving bodies are only part of the whole truth (to which Albert Einstein made an enormous contribution earlier in the twentieth century) the basic laws are still valid. They may be stated as follows.

(1) A body remains in a state of rest or of uniform motion in a straight line unless it is acted on by a resultant force.

(2) The force acting on a body is proportional to its mass and to its acceleration.

The first law gives us a definition of 'force', which is 'that which changes a body's state of rest or of uniform motion in a straight line'. The second law is used to define a unit of force.

What does the first law mean? Simply, that unless a resultant force acts on a body it will do one of two things—stay at rest or move at constant velocity along a straight line.

There is no acceleration without a resultant force.

The word *resultant* is needed because more than one force may act on a body but they may be such that their effects cancel and the result is the same as no force acting. Later we shall look more closely at getting a body to change its state from one of rest to one of motion; for the moment we have the simple truth that unless a force acts a body's state of rest or motion remains the same. The velocity does not change—note the word *velocity*, this means that the *speed* does not change and the *direction* does not change—unless a resultant force exists.

The second law, which is used to define the force unit in the SI System, indicates two things. Firstly, if it is required to give a body a particular value of acceleration then the greater the mass of the body the greater the value of the force required. Secondly, if it is required to give a body of fixed mass an acceleration then the greater the acceleration required the greater the value of the force required.

The SI unit of force is defined as that force which gives a body of mass one kilogram an acceleration of one metre per second per second. The name of the unit, appropriately enough, is the *newton*, symbol N. Using SI units then

$$\text{force} = \text{mass} \times \text{acceleration}$$

This is true only because the force unit is *defined* in the way stated. It is not always true, as in the Imperial System of units or the Metric Gravitational System of units, for instance.

Example 2.13

Find the acceleration of a body of mass 8 kg when a resultant force of 9 N is acting on it.
Solution

$$\text{Force} = \text{mass} \times \text{acceleration}$$

$$\text{acceleration} = \frac{\text{force}}{\text{mass}}$$

$$= \frac{9}{8}$$

$$= 1.125 \text{ m/s}^2$$

Example 2.14

Determine the mass of a body which, when a resultant force of 13 kN acts on it, accelerates at 1.2 m/s².

Solution

$$\text{Mass} = \frac{\text{force}}{\text{acceleration}}$$

$$= \frac{13\,000}{1.2}$$

$$= 10\,833.33 \text{ kg}$$

$$= 10.83 \text{ tonne}$$

1 tonne (unit symbol t) = 1000 kg.

GRAVITY

Most people have heard the story of the apple falling from a tree which led to Newton's so-called 'discovery' of gravity. Whether there is any truth in the tale is open to doubt; what we do know, however, is that Newton spent many years investigating what he called the gravitational attraction between bodies. He found that there is a force of attraction between all bodies, the value of the force being dependent on the masses involved. Gravity, which is usually taken to mean the force of attraction between our planet and all things on it, is one example of gravitation in general. Investigation of the force of gravity has shown that all bodies, if allowed to fall freely to the earth in any one area, are accelerated with *the same value of acceleration*. This value is constant at any one place but varies slightly at different places throughout the world, depending on the nature of the terrain (oceans, mountains, etc.). On other planets, of course, the acceleration due to the planet's gravitational pull may be vastly different. It is known that on the Moon the value is one-sixth of that on Earth (due to the Moon's smaller mass). The acceleration due to gravity is given the symbol g and its value is around 9.81 m/s^2, depending on where it is measured.

Example 2.15

A body of mass 20 kg falls freely to the earth with an acceleration of 9.81 m/s^2. Find the force of gravity acting on the body.

Solution

$$\text{Force} = \text{mass} \times \text{acceleration}$$

$$= 20 \times 9.81$$

$$= 196.2 \text{ N}$$

The force of gravity is called the *weight* of the body. Thus

weight of a body = mass × acceleration due to gravity

For a body of mass in kg

weight = mg newtons

(where g is in m/s^2). Clearly, the weight of a body, since it depends on the gravitational attraction, varies slightly from place to place on Earth and considerably elsewhere in the universe—being one-sixth of Earth-weight on the moon, for example.

Systems of units using a force unit based on weight (gravitational systems) are clearly not as good as absolute systems (like the SI) since the basic unit value is determined to some extent by where the measurement is being carried out.

It is very important to understand the difference between mass and weight. It has been decided by those responsible for metrication in the United Kingdom that the 'difference is unimportant as far as the layman is concerned'. An engineer is *not* a layman and must know that

the mass of a body is the amount of matter of the body and is measured in kilograms; the weight of a body is the force of gravity exerted on the body by the Earth and is measured in newtons.

Commercially it is unlikely that newtons will be used (in selling) so goods will probably be sold by the 'kilo' (kilogram). Oddly enough, however, they may very well be weighed by a machine the reading of which is determined by gravity and the scale of which may be marked in mass units.

Example 2.16

Determine the force exerted by the lift motors if a lift of mass 1.5 tonne is raised vertically upwards with uniform acceleration of 1.3 m/s² (take $g = 9.81$ m/s²);
Solution The resultant force on the lift is given by the equation

$$\text{resultant force} = \text{mass} \times \text{acceleration}$$
$$= 1.5 \times 10^3 \times 1.3 \text{ (mass in kg)}$$
$$= 1.95 \times 10^3 \text{ N}$$
$$\text{weight of lift} = 1.5 \times 10^3 \times 9.81$$
$$= 14.72 \times 10^3 \text{ N}$$

The *total* upward force exerted by the motors must be large enough to overcome the downward force (the weight) *and* provide the resultant force required for the acceleration of 1.3 m/s². Thus

$$\text{total upward force} = (14.72 \times 10^3) + (1.95 \times 10^3)$$
$$= 16.67 \times 10^3 \text{ N}$$
$$= 16.67 \text{ kN}$$

Example 2.17

A train of mass 250 tonne travelling at 90 km/h along a level track is brought to rest by a constant braking force of 275 kN. Calculate the time taken, neglecting windage and track friction effects.
Solution The retardation may be calculated using Newton's second law

$$275 \times 10^3 = 250 \times 10^3 \times \text{retardation}$$
$$\text{retardation} = \frac{275}{250}$$
$$= 1.1 \text{ m/s}^2$$

A speed of 90 km/h is 90 000/3600 m/s, that is, 25 m/s, thus

$$\text{time taken} = \frac{\text{speed}}{\text{retardation}}$$

$$= \frac{25}{1.1}$$
$$= 22.72 \text{ s}$$

Track friction and windage (the effect of the train pushing air out of the way) would reduce this time still further.

ASSESSMENT EXERCISES

Long Answer

2.1 Calculate the average speed of a car in m/s which travels the distances indicated in the times given: (a) 100 km in 1 h 20 min, (b) 100 m in 5 s, (c) 500 km in 5 h , (d) 1200 m in 73 s.

2.2 Determine the average acceleration of a car in m/s² which changes speed as shown in the times given: (a) from 14 m/s to 20 m/s in 1.2 s, (b) from 80 km/h to 110 km/h in 1 min, (c) from 40 km/h to 10 km/h in 31 s, (d) from standstill to 80 km/h in 4.1 s.

2.3 A car travels a journey of 100 km in three stages, each of which is travelled at constant speed. The first stage takes 1 h at a speed of 50 km/h, the last stage takes 6 h and is of length 480 km. The total journey takes 12 h. Calculate (a) the speed during each of the second and third stages (b) the average speed during stages one and two, during stages two and three and for the whole journey.

2.4 An aeroplane heading due north at a velocity of 300 km/h is blown off course by a wind blowing in an easterly direction at 40 km/h. Determine the resultant velocity of the aeroplane giving both magnitude and direction.

2.5 A train travels with uniform acceleration of 0.2 m/s² for 20 s and is then retarded uniformly at 0.4 m/s² for 18 s after which it is at rest. Find the initial speed of the train and its greatest speed during the journey.

2.6 A vehicle is uniformly accelerated from rest at 0.4 m/s² during which time it travels a distance of 1.2 km. Determine the time taken and the final velocity.

2.7 The initial speed of a moving body is 9.6 m/s; after being uniformly accelerated for 4 s the speed is 16.2 m/s. Find the distance covered during the period of acceleration.

2.8 Determine the acceleration of a body of mass 1 t when a force of 0.8 kN acts on it. If the initial speed of the body is 1 m/s find the final speed after 5 s of being subjected to this acceleration.

2.9 A body of mass 18 kg falls freely from rest to Earth from a height of 12 m. Find the velocity of the body just before it hits the ground. State any assumptions made. Take $g = 9.81$ m/s².

2.10 A lift of mass 2 t is raised vertically upwards with uniform acceleration of 1.5 m/s². Determine the tractive force exerted by the motors.

2.11 A train of mass 250 t is uniformly accelerated along a level track from rest to 100 km/h in 40 s. The tractive resistance is 75 N/t mass of train. Determine the tractive force required and the distance covered by the train during the period of acceleration.

2.12 Two cyclists A and B leave the same place at the same time and travel the same journey over the same distance. The ratio of the time taken by cyclist A to that taken by cyclist B is 2:3. What is the ratio of the average speed of cyclist B to the average speed of cyclist A? If the distance covered is 15 km and the sum of the average speeds is 35 km/h, find the time taken for the journey in hours.

2.13 The table of speed and time values for a particular journey is as follows

Time (s)	0	5	10	15	20	25	30
Speed (m/s)	14	18	22	28	28	20	15

(a) What is the average acceleration (i) in the first 10 s, (ii) between 15 and 20 s, (iii) in the last 5 s? (b) How long from the start of timing would it take for the speed to reach zero if the retardation of the last 5 s were to be maintained? (c) Is the acceleration between 0 and 10 s uniform?

2.14 The average speed of a journey lasting 5 h 45 min is 77 km/h. Determine the time taken to cover the same journey if the average speed were 55 km/h.

2.15 A stone is dropped down a pit and hits the bottom 5 s later. Calculate (a) the depth of the pit (b) the speed of the stone just before it hits the bottom.

2.16 A train of length 100 m starts from rest and is accelerated uniformly for a distance of 800 m after which it enters a tunnel of length 1 km. The rear of the train leaves the tunnel 3 min after the train started from rest. Calculate (a) the speed of the train in km/h when it enters the tunnel, (b) the speed of the train in km/h when the end of the train leaves the tunnel. Assume that the acceleration is maintained constant throughout the period that the train is travelling through the tunnel.

2.17 A car in motion is accelerated at 0.8 m/s² for 20 s after which its speed is 80 km/h. It then travels at constant speed for a distance of 1 km/h after which it is brought to rest by applying a constant retardation for a period of 12 s. Determine the initial speed of the car and the value of the constant retardation. By means of a speed-time graph determine also the total distance covered by the car throughout the period described.

Short Answer

2.18 Define speed.

2.19 Define acceleration.

2.20 Explain the difference between speed at any instant and average speed.

2.21 Explain the difference between acceleration at any instant and average acceleration.

2.22 State what the slope of a distance–time graph may be interpreted to be.

2.23 What information may be obtained from the slope of a speed–time graph?

2.24 What information may be obtained from the area under a speed–time graph?

2.25 Explain the difference between speed and velocity.

2.26 Explain what is meant by a 'vector quantity'.

2.27 Explain what is meant by 'free fall' under the influence of gravity.

2.28 Write down the equations relating distance, s, initial velocity, u, final velocity, v, and time, t, for a moving body subjected to a constant acceleration, a. What are the SI units for these various quantities?

2.29 Define the unit of force.

2.30 What is the relationship between force, mass and acceleration when a body of mass m is subjected to an acceleration a caused by a force F, all quantities being expressed in SI units? State the units involved.

Multiple Choice

2.31 A car travels 50 km in 1 hour 20 min. Its average speed in m/s is
   A. 10.41   B. 0.625   C. 625   D. 0.0104

2.32 A car travels at an average speed of 27.78 m/s. In 1 h the distance covered in km is
   A. 7.7   B. 100   C. 100 000   D. 1.67

2.33 Velocity is
   A. distance/time   B. distance/time in a given direction   C. rate of change of distance with respect to time   D. rate of change of distance with respect to time in a given direction

2.34 The average acceleration of a car which increases its speed from 10 km/h to 100 km/h in 40 s is
   A. 37.5 m/s^2 B. 2.25 m/s^2 C. 0.625 m/s^2 D. 2250 m/s^2

2.35 A car is accelerated uniformly from rest at 0.2 m/s^2; after 45 s its speed (in km/h) would be
   A. 9   B. 225   C. 4.44   D. 32.4

2.36 The force required to accelerate a vehicle of mass 1.1 t at constant acceleration of 0.3 m/s^2 is
   A. 0.272 N   B. 3.67 N   C. 330 N   D. 0.33 N

2.37 A constant force of 500 N acts on a body of mass 0.4 t. The acceleration is
   A. 0.8 m/s^2 B. 1.25 m/s^2 C. 200 m/s^2 D. 1250 m/s^2

2.38 A body falling freely under gravity
   A. moves with constant velocity   B. experiences the same force as all other bodies falling freely under gravity   C. is accelerated uniformly   D. travels at the same speed as all other bodies falling freely under gravity

2.39 A lift of mass 2 t is raised vertically upwards at constant acceleration of 1.2 m/s^2. The force required to provide the acceleration is
   A. 2.4 kN   B. 2 kN   C. 19.62 kN   D. 22.02 kN

2.40 The total tractive force in the question 2.39 is
   A. 2.4 kN   B. 2 kN   C. 19.62 kN   D. 22.02 kN

3 Force

OBJECTIVES

All the objectives should be understood to be prefixed by the words 'The expected learning outcome is that the student . . .'

F16 Solves problems involving forces in static equilibrium situations.
16.1 Defines a scalar quantity.
16.2 Defines a vector quantity.
16.3 States that force is a vector quantity.
16.4 Describes stable, unstable and neutral equilibrium.
16.5 Defines the moment of a force about a point.
16.6 States the 'Principle of Moments'.
16.7 Solves simple beam problems.
16.8 Defines centre of gravity.
16.9 Shows, with the aid of sketches, the position of the centre of gravity of (a) thin uniform rod, (b) rectangular lamina, (c) circular lamina.
16.10 Determines graphically the resultant of two co-planar forces acting at a point.

F17 Applies the principles of pressure in fluids.
17.1 Defines pressure.
17.2 States the units of pressure to be N/m^2 or pascals.
17.3 Calculates pressure given force and area.
17.4 States the factors which determine the pressure at any point in a fluid; fluid density, depth, g, surface pressure.
17.5 States that the pressure at any level in a liquid is equal in all directions.
17.6 States that the pressure is independent of the shape of the vessel.
17.7 States that the pressure acts in a direction normal to its containing surface.
17.8 States that the pressure due to a column of liquid depends on the density of the liquid and the height of the column.
17.9 Solves simple problems using $p = hdg$.
17.10 Measures gas pressure using (a) a U-tube manometer, (b) a pressure gauge.
17.11 States that there is a pressure due to the atmosphere.

E15 Describes and calculates the friction force between two surfaces in contact.
15.1 Defines friction force with the aid of a diagram.
15.2 States the factors affecting friction force size and direction.
15.3 States examples of (a) practical applications and (b) design implications of friction forces.
15.4 Solves simple problems involving use of the equation $F = \mu N$.

A body is normally in one of two states: it is either at rest or in motion. If it is in motion its speed (the rate of change of distance of the body from some fixed point with time) is either constant or is changing. If a body changes its state of rest or of uniform (constant) motion in a straight line we say that a *force* is acting on it.

Force is that which changes or tends to change a body's state of rest or of uniform motion in a straight line. There is always a direction associated with a force—the final effect of any force depends not only on its size (magnitude) but also on the direction in which it is acting. The direction in which it is acting is called the *line of action* of the force. Because force has both magnitude and direction it is called a *vector* quantity, as opposed to a *scalar* quantity which has only magnitude; examples of scalar quantities include mass and speed. Later we shall look at ways of representing vector quantities in diagram form; first we must examine how force is defined.

Many of our ideas concerning force and motion originated with the work of Isaac Newton, an English scientist who lived some 300 years ago. He spent a long time studying motion and the causes of motion, particularly the motion of the Earth and other planets relative to one another. All moving bodies have a property called *momentum*; the momentum of any moving body is proportional to the mass of the body and to the velocity of the body (velocity is speed in a given direction; velocity is a vector quantity whereas speed is a scalar quantity). Thus the momentum of a body at rest is zero and if the momentum of a body is changing either its mass is changing or its velocity is changing (or both). Newton showed that the force acting on a body produces a change in momentum and, further, that the magnitude of the force is proportional to the rate of change with time of the momentum of the body. Thus

force is proportional to the rate of change of momentum
force is proportional to the rate of change of mass × velocity

and if the mass is constant

force is proportional to mass × the rate of change of velocity

Now the rate of change of velocity is called *acceleration*, so that for a body of constant mass changing its state of rest or of uniform motion in a straight line, that is, being accelerated, the force causing the change is proportional to the product of mass and acceleration.

In the SI System of units the unit of force is taken as that force which causes a body of mass one kilogram to be accelerated at one metre per second per second, thus

force = mass × acceleration

and the unit of force is called the *newton*, symbol N. As with other units there are also multiples and sub-multiples: millinewton (mN), meganewton (MN), etc.

WEIGHT

Another aspect of motion and the causes of motion studied by Newton was the force exerted on a body due to other bodies nearby. This force, called a *gravitational* force, exists between all bodies, its size depending on the masses of the bodies concerned. The gravitational force of which we are particularly aware is that exerted by the Earth on all other bodies. This force we know as *gravity*; we are particularly aware of it because of its size (the Earth has a large mass) and because of its effects—the more so now that we are familiar with what happens when people are subject to zero gravity and excess gravity in vehicles leaving the Earth's surface and orbiting the Earth. We now know that the force of gravity at any one point on the surface of the Earth attracting other bodies to the surface will give them the same acceleration regardless of their mass. The acceleration due to gravity at any one point on the Earth is the same for all bodies. Its value varies slightly from place to place—over high mountains or deep oceans, for instance—but in any one place it is constant, the value usually being taken as 9.81 m/s^2. If force is proportional to mass × acceleration and acceleration due to gravity is constant, then the force due to gravity must be proportional to the mass alone. This force we call *weight*. Thus

the weight of a body = its mass × acceleration due to gravity

and is expressed in newtons. It is *not* correct to refer to the *weight* of a body in kilograms (or the popular abbreviation 'kilos').

Example 3.1

A body of mass 4 kg is accelerating at 3 m/s². Determine (a) the force causing the acceleration, (b) the weight of the body. Take the acceleration due to gravity, $g = 9.81$ m/s².
Solution

(a)   Force causing acceleration $=$ mass $\times$ acceleration
$$= 4 \times 3$$
$$= 12 \text{ N}$$

(b)   Weight of body (force $=$ mass $\times$ acceleration due to
on body due to gravity)   gravity

$$= 4 \times 9.81$$
$$= 39.24 \text{ N}$$

VECTOR DIAGRAMS

Force has both magnitude and direction and is therefore a vector quantity. A quantity which has only magnitude is called a scalar quantity. A scalar quantity may be represented in a diagram by marking off a length along a line using an appropriate *scale*, but the position of the line in space, that is, on the paper, is unimportant. For instance, if we wish to represent various masses, say, 1, 2, 3 and 4 kg respectively in a diagram we may take lines of length 1, 2, 3, 4 cm respectively. Here the scale is 1 cm representing 1 kg, written 'let 1 cm = 1 kg' (the equals sign means *represent* in such a case—it does *not* mean is *equal to* as is usual). The position of the line representing the various masses does not matter. An example of the kind of representation described here occurs in graph-plotting, where the way in which the value of one quantity affects the value of another is shown in diagram form.

A vector diagram uses lines drawn to scale to represent magnitude but the position of the line on the paper *does* matter and is used to show the direction of the vector. If the vector is a force,

then the position of the line shows the line of action of the force.

Figure 3.1 is in three parts. Figure 3.1a shows two forces, one of magnitude 4 N, the other of magnitude 5 N, both forces acting in the same straight line and in the same direction. Figure 3.1b shows the same two forces acting in the same straight line but this time in opposite directions. Figure 3.1c shows the same two forces but this time they do not act in the same straight line—their lines of action are separated by an angle of 50°. To draw scale drawings of these three situations we first choose a suitable scale, for example, letting 1 cm represent a force of 1 N, so that 4 N is represented by a line of length 4 cm and 5 N is represented by 5 cm. Lines of the correct length are then drawn *in a direction showing the lines of action of the forces represented by the lines*. These diagrams are vector diagrams.

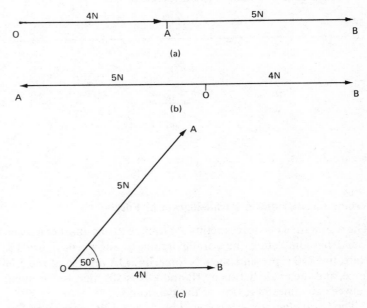

Figure 3.1

RESULTANT FORCE

If the same over-all effect of a number of forces acting on a body can be obtained with a single force, then the single force is called the *resultant* force of the number of forces it replaces. For example, the over-all effect of the two forces shown in figure 3.1a can be obtained by a single force of $(5+4)$ newtons acting along the same straight line and in the same direction as the two individual forces. The resultant force is then as shown in figure 3.2a, and has the value 9 N. The situation shown in figure 3.1b is slightly different: the two forces are opposing one another, the 5 N force acting to the left in the diagram and the 4 N force acting to the right. The same over-all effect would be obtained by a single force of 1 N acting to the left, as shown in figure 3.2b. The situation shown in figure 3.1c is different again; this time the magnitude and direction of the resultant are not immediately obvious. There are a number of ways of finding the resultant, some of which are considered below.

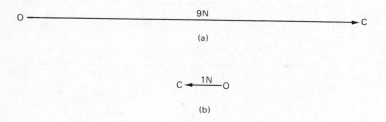

(a)

(b)

Figure 3.2

Vector Parallelogram; Parallelogram of Forces

The resultant of two forces acting at an angle to one another may be found by completing the parallelogram, as shown in figure 3.3. Here, line OB represents the 4 N force, line OA represents the 5 N force, and the angle between OB and OA is 50°, that is, the angle between the lines of action of the two forces.

The parallelogram is completed by drawing BC parallel to OA and AC parallel to OB. Line OC then represents the resultant of the two forces both in magnitude and direction. Drawing the diagram to scale, using the scale discussed earlier (1 cm = 1 N) gives length OC as 7.4 cm, which indicates a force of 7.4 N, at an angle of 31° to OB. The resultant of the two forces is then a single force of 7.4 N

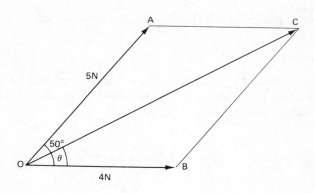

Figure 3.3

acting at an angle of 31° to the 4 N force and, since the angle between the original forces is 50°, the resultant acts at an angle of $50-31$, that is, 19° to the 5 N force.

Vector Triangle; Triangle of Forces

The resultant of two forces acting at an angle to one another may be found by constructing a triangle, as shown in figure 3.4, in which two sides represent the two forces, the third side representing the resultant of the two forces. In figure 3.4 line OB represents the 4 N

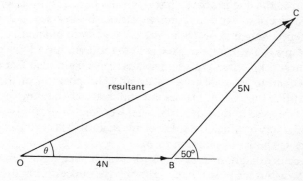

Figure 3.4

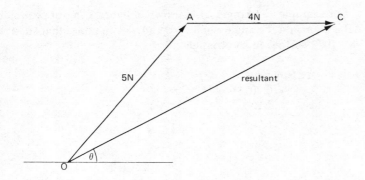

Figure 3.5

force and line BC represents the 5 N force, line OC representing the resultant both in magnitude and direction. In this method a line is drawn representing one of the forces, the other line is then drawn *from the end of the first line* to represent the other force. The resultant is then represented by the line which joins the start of the first line to the finish of the second line. The order in which the forces are taken does not matter—see figure 3.5. In this figure line OA represents the 5 N force, line AC represents the 4 N force and line OC represents the resultant. It differs from figure 3.4 in that the 5 N force was taken first. In both cases the resultant force is 7.4 N acting at 31° to the line of action of the 4 N force. As can be seen, the triangle of forces is actually one-half of the parallelogram of forces.

Both methods described above are methods involving construction of scale drawings. This is not absolutely necessary if we have a basic knowledge of trigonometry. To use trigonometry we first produce the drawing—either parallelogram or triangle—but not to scale; then we solve the triangle or parallelogram using standard methods (the sine rule or cosine rule). Trigonometric calculation of the above problem shows the resultant to be of magnitude 7.34 N acting at an angle 31° 28′ to the line of action of the 4 N force. As we can see, trigonometric calculation produces a more accurate answer than scale drawing. The use of trigonometry in the solution of vector diagrams will not be considered further at the moment except in as much as the basic sine and cosine definitions are used in vector resolution.

VECTOR RESOLUTION

As we saw above, if the forces of which we are trying to find the resultant act in the same straight line, then finding the resultant is a simple matter of addition or subtraction, depending on whether the forces act in the same direction or in opposite directions. Vector resolution is a method in which a single vector is replaced by two vectors in directions such that the simple addition or subtraction used earlier can be applied.

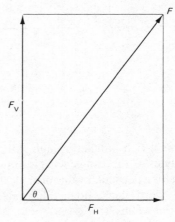

Figure 3.6

Consider two forces acting at 90° to each other, shown as F_V, drawn vertically, and F_H, drawn horizontally in figure 3.6. The resultant of these forces, using either the parallelogram or triangle method, is shown as F acting at an angle θ (pronounced 'theta') to F_H. Force F may be considered to have two *components*—one horizontal, shown as F_H, and one vertical, shown as F_V. (Horizontal means 'acting horizontally' and vertical means 'acting vertically'.) Since the two forces F_V and F_H can be replaced by resultant F, so F in turn can be replaced by F_V and F_H. This is the principle of vector resolution. Now let us examine the basic trigonometric identities used in such resolution of vectors.

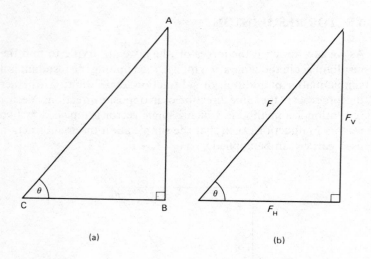

Figure 3.7

Figure 3.7a shows a right-angled triangle ABC in which angle ABC is the right-angle (90°) and angle ACB is shown as θ. The ratio AB:AC is constant for a particular value of angle θ and is called the *sine* of angle θ. (The word 'sine' is often abbreviated to 'sin', so that sine θ is written sin θ.) Thus

$$\sin \theta = \frac{AB}{AC}$$

The ratio BC:AC is also constant for a particular value of angle θ and this ratio is called the *cosine* of angle θ. ('Cosine' θ is abbreviated to 'cos' θ.) Thus

$$\cos \theta = \frac{BC}{AC}$$

There are available standard tables of values of sines and cosines of different angles so we can find ratios in triangles containing them. Also, if we know the actual length of one side of a triangle and an angle other than the right-angle, we can find the actual length of the other sides.

For example, in triangle ABC shown in figure 3.7a, suppose side AB is 5 cm and the angle is 45°. We know from tables that sin 45° = 0.7071 so that in the triangle

$$\begin{aligned} \sin ACB &= \sin 45^\circ \\ &= 0.7071 \\ &= \frac{AB}{AC} \\ &= \frac{5}{AC} \end{aligned}$$

Therefore, since 0.7071 = 5/AC then AC = 5/0.7071, which is 7.07. Thus the length of AC is 7.07 cm.

Now consider figure 3.7b, which shows the triangle of forces from figure 3.6. Here

$$\sin \theta = \frac{F_V}{F} \quad \text{and} \quad \cos \theta = \frac{F_H}{F}$$

which tells us that the vertical component of F

$$F_V = F \sin \theta$$

and the horizontal component of F

$$F_H = F \cos \theta$$

The third method of finding the resultant of two forces, then, is by resolution, which means replacing each force by its horizontal and vertical components and then finding the resultant horizontal component and the resultant vertical component (by simple addition and subtraction) and finally the single force from these resultant horizontal and vertical components. The following examples demonstrate all three methods and should be studied carefully.

Example 3.2

Find the resultant of two forces, one of 120 N acting horizontally, the other of 100 N acting at an angle of 60° to the first.

Solution Any of the three methods may be used; we shall use the parallelogram of forces, but the triangle will also be shown. The vector diagrams are shown in figure 3.8. Completion of the parallelogram in figure 3.8a gives the resultant as R at an angle θ. The triangle of forces is shown in figure 3.8b.

Scale drawing gives the magnitude of R as 191 N at an angle of 27° to the 120 N force.

The resultant of the two given forces is therefore 191 N at 27° to the horizontal along which the 120 N force is acting.

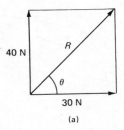
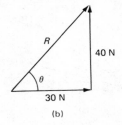

(a)   (b)

Figure 3.9

$$R = 50$$

From simple trigonometry

$$\sin \theta = \frac{40}{50}$$
$$= 0.8$$

hence

$$\theta = 53°8'$$

As before, we see that a calculation gives a rather more accurate answer than a scale drawing.

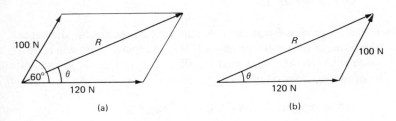

(a)   (b)

Figure 3.8

Example 3.3

Find the resultant of two forces acting at right-angles to each other, one of 40 N, the other of 30 N.

Solution The parallelogram and triangle of forces are shown in figure 3.9. Scale drawing gives the resultant as 50 N at an angle of 53° to the 30 N force.

In this particular case, because the forces are acting at right-angles to each other and the triangle of forces is a right-angled triangle, the resultant may be calculated fairly easily using Pythagoras' theorem, and the angle of action may be calculated using simple trigonometry.

In the triangle shown in figure 3.9

$$R^2 = 40^2 + 30^2$$

from Pythagoras' theorem, thus

Example 3.4

A force of 100 N acts at an angle of 15° to the horizontal. A second force of 110 N acts at an angle of 20° to the first and 35° to the horizontal. Find the resultant of these forces (a) using the parallelogram of forces and (b) by resolution.

Solution (a) The vector diagram is shown in figure 3.10a, the parallelogram of forces in figure 3.10b and the triangle of forces in figure 3.10c. (The triangle of forces is not asked for in the question but is included for demonstration.) The resultant of the two forces is shown as R in the figure, acting at an angle θ to the horizontal. Scale drawing gives the value of the resultant as 207 N acting at 25° to the horizontal.

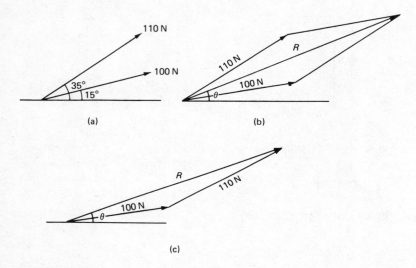

Figure 3.10

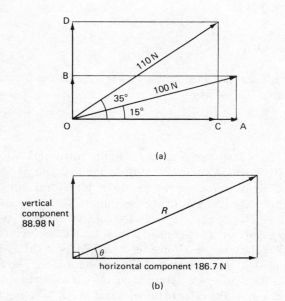

Figure 3.11

(b) The diagrams showing resolution of the two forces are contained in figure 3.11. In figure 3.11a, line OA represents the horizontal component of the 100 N force and line OC represents the horizontal component of the 110 N force. Line OB represents the vertical component of the 100 N force and line OD represents the vertical component of the 110 N force.

Using basic trigonometry

$$OA = 100 \cos 15°$$
$$OC = 110 \cos 35°$$
$$OB = 100 \sin 15°$$
$$OD = 110 \sin 35°$$

The forces represented by OA and OC are acting in the same direction, so the single horizontal force to replace them will have a value equal to their sum, that is, $100 \cos 15° + 110 \cos 35°$ which, from tables, is equal to

$$(100 \times 0.9659) + (110 \times 0.8192) = 186.7 \text{ N}$$

Similarly the vertical components may be replaced by a single force equal to their sum, since they act in the same direction, which is

$$100 \sin 15° + 110 \sin 35° = (100 \times 0.2588) + (110 \times 0.5736)$$
$$= 88.98 \text{ N}$$

So the resultant of the two forces given in the problem has a vertical component of 88.98 N and a horizontal component of 186.7 N, as shown in figure 3.11b. Applying Pythagoras' theorem to the triangle shown there

$$R^2 = 186.7^2 + 88.98^2$$

hence

$$R = 206.82 \text{ N}$$

From basic trigonometry

$$\sin \theta = \frac{88.98}{206.82}$$
$$= 0.4304$$
$$\theta = 25°29'$$

As before, calculation gives a more accurate answer.

In this last calculation the value of R was used to find the angle. To avoid the possibility of making a second error, if a first error had been made in calculating R, the *tangent* function of the angle may be used. In the triangle, tangent θ is given by dividing the vertical component by the horizontal component. Tangent functions are listed in tables with sine and cosine functions.

Vector Subtraction

Occasionally it is necessary to find the *difference* between vectors as opposed to the sum. In this case the vector to be subtracted is drawn in the *opposite* direction to that in which it acts and the vector sum of the first vector and the redrawn vector may then be found by one of the methods given. This is shown in figure 3.12, which shows two vectors, F_1 acting horizontally and F_2 acting at an angle to F_1. The vector sum of these two vectors is R, acting at angle θ to F_1 as determined by the parallelogram shown. To obtain the vector difference $F_1 - F_2$, we draw F_2 in the opposite direction, shown as $-F_2$ and obtain the sum of F_1 and $-F_2$ to give resultant R' acting at angle θ' as shown.

To obtain $F_2 - F_1$, the vector F_1 would be drawn in the opposite direction (horizontally to the left) to give $-F_1$ and this would be summed with F_2 to give $F_2 - F_1$.

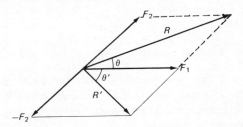

Figure 3.12

EQUILIBRIUM

If the resultant force of any system of forces is zero then the body or bodies on which the forces act is said to be in *equilibrium*. Since there is no resultant force (and, by definition, force is that which changes or tends to change a body's state of rest or of uniform motion in a straight line) then the body will carry on doing whatever it was doing before the forces were applied. If it was at rest it will remain at rest and if it was moving at constant velocity it will carry on doing so. If the resultant does have a value, then to place the body in a state of equilibrium requires a force *equal and opposite* to the resultant, so that the over-all resultant is then zero. The equal and opposite force is called the *equilibrant*.

Examples 3.5

Find the equilibrant of the forces shown in figure 3.10a.
Solution From example 3.4, the *resultant* is 206.82 N acting at 25° 29′ to the horizontal, as shown in figure 3.11b. The equilibrant is therefore 206.82 N acting in the *opposite* direction, that is, 25° 29′ − 180°, which is − 154° 31′, or 154°31′ from the horizontal in a *clockwise* direction. (Positive angles are normally measured in an anticlockwise direction.)

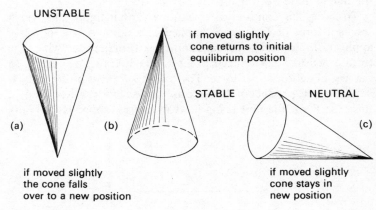

UNSTABLE

if moved slightly cone returns to initial equilibrium position

STABLE   NEUTRAL

(a)   (b)   (c)

if moved slightly the cone falls over to a new position

if moved slightly cone stays in new position

Figure 3.13

There are three kinds of equilibrium. A body is in *stable* equilibrium if, when slightly displaced from its position, it returns again to that position. If, when displaced slightly, it moves to another equilibrium position, the original equilibrium is called *unstable*. If, on moving the body, it remains in equilibrium in its new position, the original equilibrium is said to be neutral. See figure 3.13.

THE TURNING EFFECT OF A FORCE—MOMENTS

In many engineering applications we are concerned with forces which are being used for turning. An example in everyday use is the spanner or wrench used for tightening or untightening nuts. Here a force is applied to the end of the spanner, the other end being attached to the nut. The longer the spanner the less effort is required to carry out the job. Other examples include a variety of levers and bars in machinery which transmit a rotary movement from one part of the machine to another.

The turning effect of a force depends not only on the size of the force but also on the distance of the point of application of the force from the point where the turning action is required. This distance is measured along a line joining the point of application to the point where the turning action is required and situated at right-angles to the line of action of the force.

To clarify this, consider the situation shown in figure 3.14, which shows a force of F newtons applied at a point A (the point of application) situated at a distance r metres from the point where the turning action is experienced, point P. Line PA is at right-angles to the line of action of the force. The turning *moment* of the force is defined as the product of the force and the distance as described, in this case $F \times r$, the unit being newton-metres (abbreviated N m).

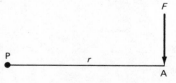

Figure 3.14

Turning moment is also described as *torque* although, strictly speaking, torque is a resultant turning moment produced by two turning forces—this will be discussed in more detail later.

Now consider a plank resting on a pivot, or *fulcrum*, as shown in figure 3.15. Suppose two forces are acting on the plank, F_1 at a distance x_1 metres from the fulcrum and F_2 at a distance x_2 metres from the fulcrum, as shown in the figure. (During the following discussion we shall ignore the weight of the plank and assume that the only forces involved in turning the plank are F_1 and F_2 as shown.)

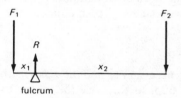

Figure 3.15

Before we examine turning moments let us first look at the forces involved. If the plank is to remain at rest, that is, move neither up nor down, then the upward forces must equal the downward forces. The total downward force is $F_1 + F_2$. The fulcrum must therefore exert a force acting vertically upwards equal to $F_1 + F_2$; this force is called the *reaction* of the fulcrum to the plank. If the plank is not to turn then the clockwise turning moment must be equal to the anticlockwise turning moment. The anticlockwise turning moment is equal to $F_1 \times x_1$ and the clockwise turning moment is equal to $F_2 \times x_2$. Thus

$$F_1 x_1 = F_2 x_2$$

and this is a necessary condition for the plank to remain in equilibrium. The reaction of the fulcrum, shown as R in the figure, does not, of course, exert a turning moment about the fulcrum since it acts *through* the fulcrum.

In general: a body free to rotate about a fulcrum is in equilibrium only if there is no resultant force acting on it and if the total clockwise moment about the fulcrum is equal to the total anticlockwise moment about the fulcrum. This statement is the *principle of moments.*

Example 3.6

A plank is balanced about a fulcrum; a force of 10 N acts vertically downwards at a distance of 0.5 m to the left of the fulcrum, and a force of 15 N acts vertically downwards on the plank to the right of the fulcrum. Ignoring the effect of the weight of the plank find the distance between the fulcrum and the 15 N force.

Solution   The situation is similar to that shown in figure 3.15, where $F_1 = 10$ N, $x_1 = 0.5$ m and $F_2 = 15$ N.

Taking moments about the fulcrum

$$10 \times 0.5 = 15 \times l$$

where l is the required distance. Hence

$$l = \frac{10 \times 0.5}{15}$$
$$= 0.33 \text{ m}$$

Thus the 15 N force acts at a distance of 0.33 m from the fulcrum.

Example 3.7

A beam of weight 25 N is in equilibrium resting on a fulcrum such that the weight of the beam to the left of the fulcrum is 10 N and the weight of the beam to the right of the fulcrum is 15 N. A force of 100 N acting vertically downwards has its point of application on the beam end at a distance of 2 m to the left of the fulcrum. Determine the distance between the fulcrum and the point of application of a 60 N force acting to the right of the fulcrum, at the other end of the beam.

Solution   Assuming a uniform beam, that is, that the beam has the same cross-section throughout and is of the same material with the same density throughout, we can say that the weight of the beam to the left of the fulcrum acts midway between the beam end and the

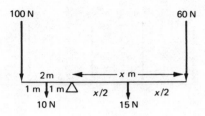

Figure 3.16

fulcrum, that is at a distance of 1 m from the fulcrum. If we let the required distance be x metres then we can assume also that the weight of the beam to the right of the fulcrum acts midway between beam end and fulcrum, that is, at a distance of $x/2$ from the fulcrum to the right (see figure 3.16).

Taking clockwise moments about the fulcrum

$$15 \times \tfrac{1}{2}x + 60 \times x = 67.5x$$

Taking anticlockwise moments about the fulcrum

$$10 \times 1 + 100 \times 2 = 210$$

Hence

$$67.5x = 210$$

and

$$x = 3.11 \text{ m}$$

Example 3.8

A beam is supported at its ends A and B as shown in figure 3.17. Forces act downwards on the beam as shown in the figure. Determine the reactions at A and B.

Solution   One technique with this type of problem is to take moments about each end in turn. In this way we obtain two

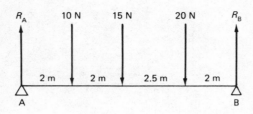

Figure 3.17

equations, each containing only one unknown quantity, since the reaction force acting through a point will have no turning moment about that point. Care should be taken when taking moments to use the correct distance for each force. In the example given, *from point A* the point of application of the 10 N force is 2 m, of the 15 N force is 4 m, of the 20 N force is 6.5 m and of force R_B is 8.5 m. *From point B* the point of application of the 10 N force is 6.5 m, of the 15 N force is 4.5 m, of the 20 N force is 2 m and of force R_A is 8.5 m.

Taking moments about point A

clockwise moments $= (10 \times 2) + (15 \times 4) + (20 \times 6.5)$
anticlockwise moments $= R_B \times 8.5$

Thus

$$(10 \times 2) + (15 \times 4) + (20 \times 6.5) = 8.5 R_B$$

and therefore

$$210 = 8.5 R_B$$

so that

$$R_B = 24.7 \text{ N}$$

Similarly, taking moments about B

$$8.5 R_A = (10 \times 6.5) + (15 \times 4.5) + (20 \times 2)$$
$$= 172.5$$

and

$$R_A = 20.3 \text{ N}$$

An alternative method is to take moments about one end, and having determined the value of one of the reaction forces, the other may be found from the fact that since the beam is in equilibrium the total upward force is equal to the total downward force; thus

$$R_A + R_B = 10 + 15 + 20$$
$$= 45 \text{ N}$$

so that

$$R_A = 45 - 24.7$$
$$= 20.3 \text{ N as before}$$

Note that the weight of the beam has not been taken into account in this example since it was not given in the problem. If the weight is given, the turning moment caused by the weight must be taken into account when taking moments.

TORQUE

The torque exerted on a lever, beam or other suitably pivoted arm is the total turning moment caused by a *couple*. A couple, which has a special meaning in this application, consists of two equal forces acting on an arm which is pivoted between the points of application as shown in figure 3.18. Taking moments about the pivot in this diagram, the clockwise moment is $Fd/2$ and the anticlockwise moment is also $Fd/2$, if the pivot is midway between the points of application of the forces. The total moment, or torque, is thus Fd. If the pivot is moved to some point a distance r from one end, as shown, the clockwise moment is now $F(d-r)$ about the pivot, and the anticlockwise moment is Fr; thus the total moment is $F(d-r) + Fr$, which equals Fd as before. The torque remains the same regardless of the position of the pivot.

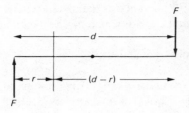

Figure 3.18

Example 3.9

A rectangular coil is free to rotate about an axis as shown in figure 3.19 and is situated in a magnetic field. When a current flows in the coil the force set up between the magnetic field of the coil and the field in which it is situated is 150 N, acting as shown. The width of the coil rectangle is 0.9 m. Determine the torque on the coil.

Solution

$$\text{Total torque} = 150 \times 0.9$$
$$= 135 \text{ N m}$$

(This example illustrates a basic electric motor.)

particles joined together. The force of gravity acts on each of these particles and the resultant of all these forces, which may be assumed to act in directions parallel to one another, is the total force of gravity on the body, that is, the weight of the body. The point through which the weight of a body acts is called the *centre of gravity* of the body. If moments are taken about the line of action of the weight through the centre of gravity the resultant moment is zero (clockwise moments = anticlockwise moments) and the body is in equilibrium. If a body, a lever or bar, for example, is pivoted at its centre of gravity it will balance.

For bodies with regular shapes the position of the centre of gravity may be found easily by inspection. For example, as we have seen in some of the examples given earlier, a beam of constant cross-section and uniform throughout (being made of the same material throughout) will have its centre of gravity at the beam centre. Similarly, a uniform sphere or ball has its centre of gravity at the centre. In cases where uniformity is not present or where shapes are irregular, other methods must be used. These include experimental methods, methods involving taking moments and mathematical methods using calculus. (We shall not be considering this last method.)

A useful experimental method is illustrated in figure 3.20. Here the centre of gravity of a *lamina* (meaning a thin layer) of triangular shape is being found. The shape is suspended to hang freely under

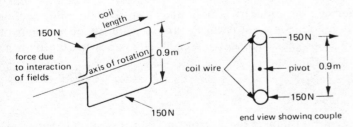

Figure 3.19

CENTRE OF GRAVITY

Any body may be considered to be made up of a large number of

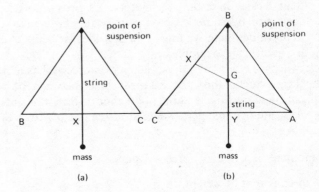

Figure 3.20

gravity. A string, to which is attached a suitable mass, is also suspended from the same point. In the diagram the lamina is suspended at A, as is the string, which hangs in front of the lamina. The point where the string passes across line BC is marked on the lamina, point X. If the lamina is hanging freely and at rest then moments taken about any point on line AX are equal. The centre of gravity lies somewhere along this line. The lamina is removed and line AX drawn on it. The process is then repeated with the lamina suspended from another point, shown as B in the diagram. Line BY is drawn and where the lines cross, point G in figure 3.20, is the centre of gravity of the lamina. Point G is referred to as the *centroid* or *centre of area* if we are considering just the area of the lamina. The centre of area of any given surface is the point at which the centre of gravity would lie if the surface were that of a lamina.

Centres of gravity may also be found by calculation, by taking moments about an axis. If the body consists of a number of regular shapes joined together and the centre of gravity position is known for each part, then, since the sum of the moments of each part about any axis is equal to the total moment, we can say that

the sum of the products of the weight of each part and the distance of the centre of gravity of each part from the axis is equal to the weight of the whole body multiplied by the distance between the centre of gravity of the whole body from the axis about which moments are being taken.

As is often the case, describing the method in words is not as clear as demonstration by example. Some examples are given below, but before we examine them we must first consider *moments of area*, which are used in the examples.

So far we have considered the moment of a force and have taken it to be the product of the force and the distance between the point of application of the force and the point about which moments are being taken, along a line perpendicular to the line of action of the force. Thus

$$\text{moment of a force} = \text{force} \times \text{distance}$$

If the force is that of weight, then

$$\text{moment of weight} = \text{weight} \times \text{distance}$$

and in this case the 'distance' is the distance between the centre of gravity and the point about which moments are being taken. Now

$$\text{weight} = \text{volume} \times \text{density}$$

where 'density' means the weight per unit volume of the body material, and, for a body of uniform thickness

$$\text{volume} = \text{area} \times \text{thickness}$$

so that

$$\begin{aligned}\text{moment of weight} &= \text{area} \times \text{thickness} \times \text{density} \times \text{distance} \\ &= (\text{area} \times \text{distance}) \times \text{density} \\ &\quad \times \text{thickness}\end{aligned}$$

The bracketed quantity, area × distance, which means the surface area × the distance between the point about which moments are being taken and the centre of gravity (or, more specifically, since we are considering an area—the centroid) is called the *moment of area*. Thus

$$\text{moment of weight} = \text{moment of area} \times \text{density} \times \text{thickness}$$

and for the same body, when we are using the fact that the sum of moments of various parts of the body is equal to the moment of the whole body about any particular axis or point, we can use moments of area instead of moments of weight. This is shown in the following examples.

Example 3.10

Find the position of the centroid of the area shown in figure 3.21a.
Solution The area consists of two rectangles shown as CJKH and DEFG. The first thing to do is to locate the position of the centroid of each of these parts. The centroid of a rectangle lies at the point of intersection of the diagonals, so that the centroid of rectangle CJKH lies at G_1 and that of rectangle DEFG lies at G_2 as shown in figure 3.21b. By inspection, the distance between G_1 and side CH is

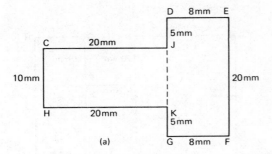

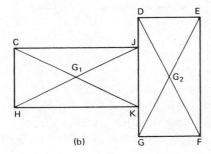

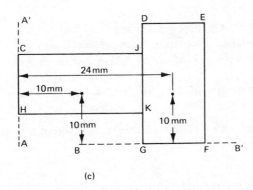

Figure 3.21

10 mm, between G_1 and side HK is 5 mm, the distance between G_2 and side DG is 4 mm, between G_2 and side FG is 10 mm.

The method now involves taking moments of area of these two rectangles about some convenient axes. These have been taken as AA′ and BB′ as shown in figure 3.21c. The distance between AA′ and G_1 is 10 mm, between G_1 and BB′ is 10 mm, between G_2 and AA′ is 24 mm and between G_2 and BB′ is 10 mm. These distances are perpendicular distances, that is, along a line between the centroids and the axes at right-angles to the axes.

The area of rectangle CJKH is $20 \times 10 = 200$ mm². The distance between its centroid G_1 and AA′ is 10 mm and between its centroid and BB′ is also 10 mm. Thus

$$\begin{aligned}
\text{moment of area of CJKH about AA′} &= \text{area} \times \text{distance} \\
&= 200 \times 10 \\
&= 2000 \\
\text{moment of area of CJKH about BB′} &= 200 \times 10 \\
&= 2000
\end{aligned}$$

The area of rectangle DEFG is $20 \times 8 = 160$ mm². The distance between its centroid G_2 and AA′ is 24 mm and between its centroid and BB′ is 10 mm. Thus

$$\begin{aligned}
\text{moment of area of DEFG about AA′} &= \text{area} \times \text{distance} \\
&= 160 \times 24 \\
&= 3840 \\
\text{moment of area of DEFG about BB′} &= 160 \times 10 \\
&= 1600
\end{aligned}$$

The total area of the surface is $200 + 160 = 360$ mm². Let the distance between the centroid of the whole area and AA′ be x and between the centroid of the whole area and BB′ be y, measured perpendicular to these axes. Then

$$\begin{aligned}
\text{moment of area of whole surface about AA′} &= 360x \\
\text{moment of area of whole surface about BB′} &= 360y
\end{aligned}$$

Hence, taking moments of area about AA′

$$360x = 2000 + 3840$$
$$= 5840$$

and

$$x = \frac{5840}{360}$$
$$= 16.2 \text{ mm}$$

And taking moments of area about BB′

$$360\, y = 2000 + 1600$$
$$= 3600$$
$$y = \frac{3600}{360}$$
$$= 10 \text{ mm}$$

Thus the centroid of the area as a whole is situated 16.2 mm from AA′ and 10 mm from BB′. (The latter distance could be obtained by inspection, since the surface is symmetrical about an axis drawn parallel to BB′ through G_1 and G_2 and moments taken about such an axis would be equal.)

Example 3.11

Find the centre of area of the surface CDEFGHJM shown in figure 3.22.

Solution The shape consists of three rectangles CDLM, EFKL and GHJK. The centroid of each of these areas is determined by inspection. The centroids are shown as G_1, G_2 and G_3. The method of solution is as for the previous example.

Area CDLM

Area $= 40 \times 50 = 2000$ mm²
Distance of centroid from AA′ = 20 mm
Moment of area about AA′ $= 2000 \times 20 = 40\,000$

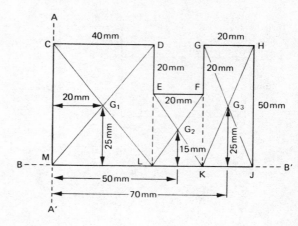

Figure 3.22

Area EFKL

Area $= 20 \times 30 = 600$ mm²
Distance of centroid from AA′ = 50 mm
Moment of area about AA′ $= 600 \times 50 = 30\,000$

Area GHJK

Area $= 20 \times 50 = 1000$ mm²
Distance of centroid from AA′ = 70 mm
Moment of area about AA′ $= 1000 \times 70 = 70\,000$

Total moment of area is then $40\,000 + 30\,000 + 70\,000$ which equals 140 000. But the total moment of area is equal to

total area × distance between centroid of total area from AA′

that is,

$(2000 + 600 + 1000)$ × required distance

which is

3600 × distance

hence

$$3600 \times \text{distance} = 140\,000$$

and distance of centroid of whole area from axis AA' is therefore

$$\frac{140\,000}{3600} = 38.89 \text{ mm}$$

Taking moments about axis BB': total area × distance of centroid of whole area from BB' is equal to moment of area of CDLM plus moment of area of EFKL plus moment of area of GHJK. Thus

$$3600 \times \text{distance} = (2000 \times 25) + (600 \times 15) + (1000 \times 25)$$
$$= 50\,000 + 9000 + 25\,000$$
$$= 84\,000$$

hence

$$\text{distance} = \frac{84\,000}{3500}$$
$$= 24 \text{ mm from axis BB'}$$

The centroid of the whole area is therefore 38.89 mm from axis AA' and 24 mm from axis BB'. These two distances locate the centroid completely. If we had a uniform lamina (of constant thickness and the same material throughout) of this shape, the centre of gravity of the lamina would be at the centroid of the area as determined in this example.

PRESSURE

When a force acts in a direction perpendicular to a particular surface the quantity obtained by dividing the force by the area of the surface is called *pressure*. The basic units are newtons per square metre, abbreviated N/m^2 and one newton per square metre is called one *pascal*, symbol Pa. (An older unit still in use is the bar, which equals 100 kN/m^2.)

Example 3.12

A force of 100 N acts uniformly over an area of 5 m^2. Calculate the pressure on the area assuming that the force acts at right-angles to the surface.
Solution

$$\text{Pressure} = \frac{\text{perpendicular force}}{\text{area}}$$
$$= \frac{100}{5} \text{ N/m}^2$$
$$= 20 \text{ Pa}$$

Example 3.13

A force of 100 N acts at an angle of 30° to a surface, the surface being a square of side 0.5 m. Calculate the pressure on the surface.
Solution

Component of force acting at right-angles to the surface $= 100 \sin 30°$
$$= 50 \text{ N}$$

Surface area $= 0.5^2$
$$= 0.25 \text{ m}^2$$

$$\text{Pressure} = \frac{50}{0.25} \text{ N/m}^2$$
$$= 200 \text{ Pa}$$

Example 3.14

A high-heeled shoe has a heel of circular cross-section of diameter 10 mm where it is in contact with the floor. If the person wearing the shoe has a mass of 52 kg and the full weight of the person is impressed on the shoe heel, calculate the pressure on the floor.
Solution

Force acting on heel $=$ weight of person
$$= 52 \times 9.81$$
$$= 510.1 \text{ N}$$

Surface area on which force acts = surface area
$$\text{of heel at floor}$$
$$= \frac{\pi \times 0.01^2}{4}$$
$$= 0.000\,078 \text{ m}^2$$

$$\text{Pressure} = \frac{\text{force}}{\text{area}}$$
$$= \frac{510.1}{0.000\,078}$$
$$= 6\,498\,344 \text{ N/m}^2$$
$$= 6.498 \text{ MN/m}^2 \text{ or MPa}$$

We see that although the force is relatively small the pressure on the floor is considerable. The reason, of course, is the small surface area over which the force acts. (Such a high pressure on certain floor surfaces can damage them, which is why shoes of small cross-sectional area at the heel should not be worn on certain floors.)

Fluid Pressure

In discussing pressure exerted on or by fluids the term 'fluid' means either a liquid or a gas. A fluid normally takes up the shape of its container as far as the volume of the fluid allows. If pressure is applied to a fluid it is transmitted equally in all directions and if a fluid is used to transmit pressure, as in a hydraulic brake system for example, the force exerted by the fluid on a surface is at right-angles to the surface and the fluid pressure is the same in all directions. The pressure at any point in a fluid depends on the depth of the point below the fluid surface, the density (mass per unit volume) of the fluid, the value of g (the acceleration due to gravity), since this affects fluid weight and whatever surface pressure is present. The pressure does not depend on the shape of the container if the fluid is a liquid, but since a gas normally expands to fill a closed container, and volume and pressure are related according to the general gas laws, pressure is affected to some extent by container shape in the case of a gas. These points will be clarified by the following examples.

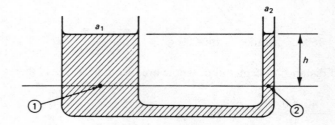

Figure 3.23

If water or some other liquid is poured into a U-tube having different diameters, as shown in figure 3.23, the level of liquid on each side will be as shown. The pressure on the liquid surfaces is that due to the atmosphere (atmospheric pressure) and is the same on each side of the tube. Consider now the pressure at points 1 and 2 in the liquid, point 1 being in the wider part of the tube, point 2 in the narrower part, both points are at the same level. The left-hand column of liquid exerts a pressure transmitted through the fluid to the right-hand column; similarly the right-hand column of liquid exerts a pressure on the left-hand column. In each case the pressure is due to the weight of the column of liquid plus atmospheric pressure (which, since it is the same in both cases, may be ignored). Since the fluid does not move, the pressure transmitted by the left-hand column at point 1 must equal the pressure at point 2 transmitted by the right-hand column. This is true wherever points 1 and 2 are, provided they are at the same level below the surface. If the pressure is the same, then what is its value? The pressure is caused by weight and the weight of either column is given by mass × acceleration due to gravity. The mass in each case is obviously different. Pressure, however, is force/area; in this case the pressure due to the liquid column is weight/area, and since the areas are different, the pressure in each case ends up the same value. If the area of either side is increased, the mass and thus the weight is increased and the pressure remains the same at each side at points at the same level. Using symbols, we denote areas by a_1 and a_2 as shown and height by h; then

weight of left-hand column $=$ mass $\times g$
$$= \text{density} \times \text{volume} \times g$$
$$= \text{density} \times a_1 hg$$

$$\text{pressure due to liquid alone} = \frac{\text{weight}}{\text{area}}$$

(neglecting atmospheric pressure)

$$= \text{density} \times a_1 hg \div a_1$$
$$= \text{density} \times h \times g$$

and, clearly, the same result will be obtained for the right-hand column

weight $=$ mass $\times g$
$$= \text{density} \times \text{volume} \times g$$
$$= \text{density} \times a_2 hg$$

pressure due to liquid alone $=$ density $\times a_2 hg \div a_2$
$$= \text{density} \times h \times g \text{ as before}$$

In general, we see the important result that the pressure at any point in a liquid depends on its density, the depth of the point below the surface and the value of g, as stated above. Note that the pressure at a point in the liquid is the sum of the pressure due to the weight of the liquid above the point and whatever pressure is being exerted on the liquid surface. The surface pressure in the example above is atmospheric pressure.

Since the pressure in a fluid at any point depends only on density, depth of point and the value of g, then pressures may be compared using only the depth figure. In this case the depth figure is called the *head* of the fluid above the point at which pressure is measured. The principle is used in various methods of measurement of pressure considered later in the section. Before that, however, some worked examples are given.

Example 3.15

A U-tube of uniform cross-sectional area containing a liquid of density 13.5 tonne/m^3 is connected at one end to a gas outlet while the other end is left open to the atmosphere. The liquid level at the open end rises to a height of 250 mm above the level of the liquid at the closed end. Calculate the pressure of the gas at the outlet if atmospheric pressure is 100 kN/m^2.

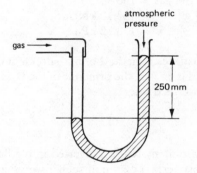

Figure 3.24

Solution The situation is shown in figure 3.24. On the left-hand side of the tube, at the level of the liquid the pressure is due to the gas from the outlet. At this level on the right-hand side of the tube the pressure is due to atmospheric pressure plus the pressure due to a 250 mm height of liquid of density 13.5 tonne/m^3. The pressures at each side *at the same level* are equal and thus if we calculate the pressure at this level at the right-hand side we are calculating the pressure of the gas from the outlet.

Weight of liquid $=$ mass $\times g$
$$= \text{volume} \times \text{density} \times g$$
$$= \text{area} \times \text{height} \times \text{density} \times g$$
Pressure due to liquid $= \dfrac{\text{weight}}{\text{area}}$
$$= \text{height} \times \text{density} \times g$$
$$= 0.25 \times 13.5 \times 10^3 \times 9.81 \text{ N/m}^2$$

taking g as 9.81 m/s². (Note that the height is in metres now and the density in kilograms/cubic metre, 13.5×10^3 kg/m³.) Hence

pressure due to liquid = 33.12 kN/m²
atmospheric pressure = 100 kN/m²

Thus

pressure of the gas = 100 + 33.12
                    = 133.12 kN/m²
                    = 133.12 kPa

The example illustrates a method of measurement of gas or other fluid pressure and is, in fact, the principle of the *U-tube manometer*, which will be described later.

Example 3.16

A test tube containing mercury is inverted into a bath of mercury exposed to atmospheric pressure in such a way that no air enters the test tube. When the inverted tube is upright the mercury inside the tube settles at a level 755 mm above the level of the mercury in the bath. The density of mercury may be taken as 13.4 tonne/m³. Calculate the pressure exerted by the atmosphere on the mercury in the bath.

Solution   The situation is shown in figure 3.25. Since a vacuum

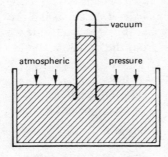

Figure 3.25

exists at the top of the inverted tube the total pressure inside the tube at the level of the mercury in the bath is due to a column of mercury 755 mm high inside the tube. Since levels are now settled, that is, equilibrium has been reached, this pressure is equal to that exerted by the atmosphere on the mercury bath. (If the tube pressure were greater, the mercury level inside the tube would fall until it equalled atmospheric pressure; if it were less the tube mercury level would rise.)

$$\text{Atmospheric pressure} = \frac{\text{pressure inside tube due}}{\text{to 755 mm of mercury}}$$
$$= 0.75 \times 13.4 \times 10^3 \times 9.81 \text{ Pa}$$

(that is, height × density × acceleration due to gravity). Thus atmospheric pressure = 0.986×10^5 Pa (N/m²)

This method of measuring atmospheric pressure is used in the *mercury barometer*. Atmospheric pressures are commonly referred to in terms of heights of mercury and the actual value in pascals may be determined using the above method.

Methods of Measuring Pressure

Two methods have already been considered; the U-tube manometer and the mercury barometer. In both cases the tube containing the fluid may be mounted against a calibrated scale so that pressures can be read off directly. Both these methods are useful for measuring pressures at or near atmospheric, although the mercury barometer (since it is not easily made portable) is usually replaced by an *aneroid* barometer, particularly for domestic use. In this instrument a sealed vessel containing a reduced air pressure is exposed to atmospheric pressure so that the vessel shape is distorted by the difference in internal and external pressure; the distortion is then arranged to produce a deflection of a pointer over a suitable scale.

Higher pressures may be measured by a *Bourdon gauge* in which the pressure being measured causes a plunger to move a pointer over a scale, the movement being amplified (made larger) by suitable small gearwheels.

Very low pressures may be measured using a *Pirani gauge* in which the variation of resistance of a heated filament due to changing thermal conductivity of a gas at different pressures is converted by electrical means to a mechanical or electrical indication of pressure.

FRICTION

When two surfaces are placed together and one is made to slide over the other, a force is set up opposing the sliding action. This force is called *friction*. There are two kinds of friction, one is called *static* friction, the other *sliding* friction. Static friction occurs when the two surfaces are initially at rest and an attempt is made to start the sliding action. To do this a force must be exerted in the proposed direction of motion to overcome the static friction. Once the sliding action begins there is still a frictional force to overcome and it is this force which is called sliding friction. Static friction, sometimes called *stiction*, is slightly greater than sliding friction.

It is found that, once sliding has begun, frictional forces are proportional to the forces at right-angles to the surfaces (weight, surface reaction, etc., depending on the situation), are independent of the area in contact and, provided sliding speeds are relatively low, are independent of sliding speed.

All surfaces rubbing together exhibit friction. If it is a problem (as in machine bearings) the effect can be substantially reduced by the use of lubricants, which are materials with an atomic structure such that the material layers slide easily over one another. When the lubricant is placed between the rubbing surfaces, which on their own do not slide easily, the over-all frictional forces between the surfaces are thus reduced.

Coefficient of Friction

Consider a body of mass m kg sliding along a surface as shown in figure 3.26. The weight of the body acts vertically downwards (assuming a horizontal surface) and if there is no resultant movement in a vertical direction (and therefore no resultant force acting) there must be a surface reaction, equal and opposite to the

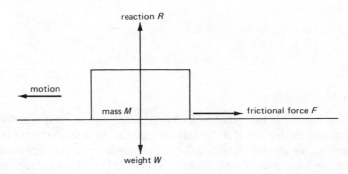

Figure 3.26

weight, acting upwards. The force causing motion acts in the direction of motion and the frictional force set up by the sliding action directly opposes the force causing motion, as shown in the figure. The ratio between the sliding friction force and the surface reaction is called the *coefficient of friction* and for two particular materials it is found to be substantially constant. The coefficient of friction is a pure number since it is a ratio of two forces. The symbol for coefficient of friction is μ (pronounced 'mu') and tables of values of μ for various situations are available.

Example 3.17

In a test on brake-lining materials it was found that to slide a 1 kg mass of the material over a horizontal metal surface a force of 5.5 N was required. Calculate the coefficient of friction for this material sliding on metal.

Solution The weight of the brake lining is mass × acceleration due to gravity, that is, 1×9.81 N (taking $g = 9.81$ m/s^2). The surface reaction is therefore 9.81 N acting vertically upwards. The horizontal force required to overcome friction is given as 5.5 N. We may assume then that the frictional force is 5.5 N (in fact, it will be slightly less since there must be a small resultant to cause motion). Thus

$$\text{coefficient of friction} = \frac{\text{frictional force}}{\text{surface reaction}}$$

$$= \frac{5.5}{9.81}$$

$$= 0.561$$

Angle of Friction

When a body is subjected to frictional force, the normal reaction force and the frictional force may be replaced by a single resultant reaction, as shown in figure 3.27. The angle that this resultant reaction force makes with the line of action of the normal reaction force is called the *angle of friction*, the symbol usually used being ϕ (phi). The tangent of this angle, from the force triangle shown in the figure, is given by

$$\tan \phi = \frac{F}{R_N}$$

On the point of movement, when the frictional force becomes that due to sliding friction

$$F = \mu R_N$$

so that

$$\tan \phi = \frac{\mu R_N}{R_N}$$

$$= \mu$$

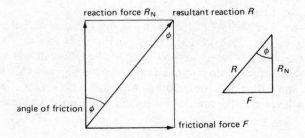

Figure 3.27

Thus *the tangent of the angle of friction is equal to the coefficient of friction.*

ASSESSMENT EXERCISES

Long Answer

3.1 State the law connecting force, mass and acceleration if the quantities are measured in SI units. A body of mass 3.7 kg is being uniformly accelerated at 0.5 m/s^2. Calculate the force causing the acceleration. If $g = 9.81$ m/s^2 find the weight of the body.

3.2 Explain what is meant by equilibrium when a system of forces acts on a body. Find the resultant force of three forces acting as follows: 45 N acting horizontally, 28 N acting at 15° to the horizontal and 35 N acting horizontally in the opposite direction to that of the 45 N force.

3.3 Find by resolution the resultant force of a system of forces acting as follows: 145 N acting horizontally to the right, 320 N acting vertically upwards, 215 N acting at 25° to the horizontal to the right upwards, 417 N acting at 15° to the vertical to the right upwards.

3.4 Find by drawing the resultant of the forces shown in figure 3.28.

3.5 Find by resolution the resultant of the forces shown in figure 3.29. (Take two forces at a time.)

3.6 State the principle of moments. A beam of negligible mass is balanced about a fulcrum; a force of 50 N acts vertically downwards at a distance of 0.7 m to the left of the fulcrum and a 30-N force acts vertically downwards to the right of the fulcrum at a distance of x m from the fulcrum. Find x.

3.7 A beam of mass 0.7 kg is in equilibrium resting on two supports. A mass of 1 kg is placed on the beam between the

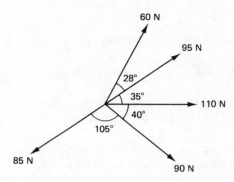

Figure 3.28

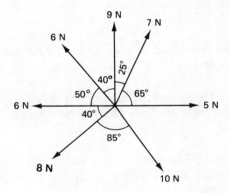

Figure 3.29

supports at a distance of 0.4 m from one end and 0.7 m from the other end. Assuming uniform mass distribution of the beam and that the supports are situated at the beam ends, calculate the reaction at each support. Take $g = 9.81$ m/s^2.

3.8 A crowbar of length 1.1 m is pivoted about a point a distance of 0.3 m from the load end of the bar. Calculate the force exerted on the load when an effort force of 150 N is exerted at the crowbar end.

3.9 Define the moment of a force about a point. A wheelbarrow is used to carry a load of weight 200 N situated 0.35 m from the fulcrum. The upward force exerted at the handle to lift the load is 52 N. Neglecting the barrow weight determine the barrow length (from fulcrum to handle).

3.10 Define the terms 'centre of gravity' and 'centroid of area'. Find the position of the centroid of the area shown in figure 3.30.

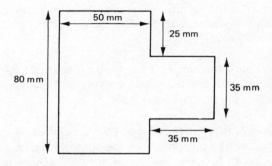

Figure 3.30

3.11 Find the centre of area of an equilateral triangle illustrating clearly the method used.

3.12 Explain the difference between a vector and a scalar quantity. A body of mass 12 kg moving at a constant speed of 7.5 m/s is acted on by a constant force of 32 N, the line of action of the force being at an angle of 40° to the direction of motion of the body. Find the velocity (magnitude and direction) after 15 s from the initial application of the force.

3.13 A horizontal beam of length 2.5 m is simply supported at each end and carries a vertical load of 250 N situated 1.8 N from one end. Calculate the reactions at the supports.

3.14 A rectangle of side lengths 50 mm and 120 mm has a square of side 25 mm removed from it. The centre of the square is equidistant from adjacent sides of the rectangle, being situated at a distance of 18 mm from one of them. Find the centre of area of the resultant shape.

3.15 A uniform beam of mass 20 kg and length 3 m is pivoted at one end and supported by a wire at the other. A 5 kg mass is placed at 0.75 m from the pivot. Find the tension in the wire and the reaction exerted by the pivot if the beam is held in equilibrium.

3.16 Calculate the height of mercury in a mercury barometer corresponding to an atmospheric pressure of 100 kPa. (Density of mercury = 13.6 t/m^3)

3.17 If the pressure due to a 1.2 m head of fluid is 8 kPa, calculate the pressure and hence the total force on a 10 mm diameter circle due to a 5-m head of the same fluid.

3.18 Calculate the total pressure at a point 1.5 m below the surface of a tank of water exposed to the atmosphere. Atmospheric pressure may be taken as 765 mm of mercury; density of mercury may be taken as 13.6 t/m^3 and density of water as 1 t/m^3.

Short Answer

3.19 A body of mass 5 kg is uniformly accelerated at 2 m/s^2. Calculate the force causing the acceleration.

3.20 A force of 2 kN acts on a body of mass 0.5 t. Calculate the acceleration.

3.21 A force of 1.5 kN causes a body to accelerate uniformly at 1.2 m/s^2. Calculate the mass of the body.

3.22 Calculate the turning moment about a point of a force of 100 N acting at a distance of 0.2 m from the point.

3.23 A force of 1 kN has a moment of 250 N m about a certain point. Calculate the distance between the point and the point of application of the force.

3.24 Calculate the weight of a body of mass 250 kg.

3.25 Calculate the mass of a body of weight 1000 N.

3.26 Calculate the pressure exerted by a force of 0.5 kN acting perpendicular to a surface of area 0.1 m^2.

3.27 A perpendicular force of 80 N sets up a pressure of 0.05 Pa when acting on a certain surface. Calculate the surface area.

3.28 Calculate the value of the perpendicular force necessary to set up a pressure of 1 Pa on a surface of area 0.2 m^2.

3.29 Calculate the pressure of a 0.5 m height of liquid of density 13.5 t/m^3.

3.30 Calculate the density of a liquid if a 0.25 m height sets up a pressure of 12.26 kPa.

3.31 Calculate the force necessary to just slide a body of mass 1 kg over a surface if the coefficient of friction between the body and the surface is 0.6.

3.32 A force of 250 N just moves a body of mass 50 kg over a horizontal surface. Calculate the coefficient of friction between body and surface.

3.33 A force of 1 kN just moves a body over a horizontal surface. The coefficient of friction between body and surface is 0.4. Calculate the mass of the body.

Multiple Choice

3.34 A force of 100 N acts on a body of mass 50 kg. Its acceleration is

  A. 5 m/s^2 B. 5000 m/s^2 C. 0.5 m/s^2 D. 2 m/s^2

3.35 A body of mass 50 kg accelerates uniformly at 4 m/s^2. The force acting is
  A. 0.08 N B. 200 N C. 200 kN D. 12.5 N

3.36 A force of 2 kN acts on a body giving it an acceleration of 0.5 m/s^2. The mass of the body is
  A. 1000 kg B. 0.25 kg C. 4 kg D. 4000 kg

3.37 A force of 0.5 kN acts on a body of mass 0.4 t. Its acceleration is
  A. 1.2 m/s^2 B. 1.2 km/s^2 C. 0.8 m/s^2 D. 0.8 km/s^2

3.38 A body of mass 1.2 t accelerates uniformly at 0.002 km/s^2. The force acting is
  A. 0.0024 kN B. 2.4 kN C. 2.4 N D. 600 N

3.39 A force of 1.8 kN acts on a body giving it an acceleration of $2 \times 10^{-3} \text{ km/s}^2$. The mass of the body is
  A. 0.9×10^{-3} t B. 3.6 t C. 0.9 t D. 3.6 kg

3.40 If a constant force is applied to bodies of different masses
  A. the size of the mass has no effect on the acceleration B. the smaller the mass the smaller the acceleration C. the larger the mass the greater the acceleration D. the larger the mass the smaller the acceleration

3.41 Select one of the following groups of quantities in which one of the members is out of place.
  A. mass, length, speed B. mass, length, time C. mass, speed, force D. force, velocity, rate of change of velocity

3.42 A system of forces consists of a force of 100 N acting to the right, one of 50 N acting to the left, one of 100 N acting vertically upwards and one of 50 N acting vertically downwards. The single resultant force to replace this system
  A. acts at 45° to the horizontal and is of magnitude 70.7 N B. is zero C. acts at 45° to the horizontal and is of magnitude 100 N D. cannot be determined without further information

3.43 A beam of negligible mass is balanced at its centre. A force acts to the left of centre in a direction perpendicular to the beam and at a distance x from the centre. If a second force acting to the right of centre, also in a direction perpendicular to the beam, has a magnitude twice that of the force acting to the left, the distance between the point of application of the force acting to the right of centre and the centre of the beam for equilibrium to be maintained
  A. is x B. is $2x$ C. is $0.5x$ D. cannot be calculated without further information.

3.44 The weight of a body of mass 0.5 t, assuming $g = 9.81 \text{ m/s}^2$, is
  A. 50.96 N B. 4.905 kN C. 4.905 kg D. 4.905 N

3.45 A force of 150 N acts uniformly over an area of 0.5 m^2. The pressure on the area, assuming the force acts at right-angles to the surface, is
  A. 300 Pa B. 75 N/m^2 C. 75 kPa D. 300 kPa

3.46 A pressure of 150 Pa is set up by a force acting perpendicular to a surface of area 0.2 m^2. The force is of magnitude
  A. 30 N B. 30 kN C. 300 N D. 0.0013 N

3.47 For a force of 0.2 kN acting perpendicular to a surface to set up a pressure of 0.3 kPa the surface area must be
  A. 1.5 m^2 B. 0.67 m^2 C. 0.06 m^2 D. 666.67 m^2

3.48 The pressure at a point 15 m below the surface of a liquid of density 13.5 t/m^3 due to the liquid alone is
  A. 202.5 Pa B. 202.5 kPa C. 1.11 kPa D. 1.99 MPa

3.49 The pressure at any point in a liquid depends solely on
  A. depth below the surface B. depth below surface, liquid density C. depth below surface, acceleration due to gravity, liquid density D. acceleration due to gravity, liquid density, depth below surface, pressure above liquid surface

3.50 In a U-tube manometer the pressure exerted by the column of liquid at the open end above the level of that at the closed end is 40 kN/m^2. The gas pressure at the closed end is 140 kN/m^2.

Atmospheric pressure is therefore equal to
  A. $0.286 \, kN/m^2$ B. $3.5 \, kN/m^2$ C. $180 \, kN/m^2$
  D. $100 \, kN/m^2$

3.51 It is found that to slide a mass of 50 kg over a horizontal surface a force of 300 N is required. The coefficient of friction between the surfaces is
  A. 0.6   B. 6   C. 1.635   D. 0.61

3.52 The coefficient of friction between two surfaces is 0.75. To just slide a 1 kg mass of the one material over the other the force necessary is
  A. 0.75 N   B. 7.36 N   C. 13.08 N   D. 1.33 N

3.53 A force of 7 kN just slides a body over a horizontal surface. The coefficient of friction is 0.5. The mass of the body is
  A. 3.5 t   B. 3.5 kg   C. 1427.1 kg   D. 14 t

4 Work and Energy

OBJECTIVES

All the objectives should be understood to be prefixed by the words
'The expected learning outcome is that the student . . .'

B4 Describes the nature and types of energy.
  4.1 Describes fuels as a source of energy.
  4.2 Describes the relationship between energy and work done.
  4.3 Defines work in terms of force applied and distance moved.
  4.4 Defines the joule.
  4.5 Draws graphs, from experimental data, of force against distance moved and relates work done with area under the graphs.
  4.6 Names five forms of energy.
  4.7 Gives two examples of conversion of heat energy to other forms of energy and vice versa.
  4.8 Defines efficiency in terms of energy input and output.
  4.9 States that power is the rate of transfer to energy.
  4.10 States that the unit of power is the watt.

WORK

When a body changes its state of rest or of uniform motion the change is caused by a *force*. Work is done whenever a force moves its point of application. The work done is proportional to the force and to the distance moved by the force. In the SI System of units the unit of work is taken as the unit of force × the unit of distance. Now the unit of force is the *newton* (symbol N) and the unit of distance is the metre, so the unit of work is the newton-metre. The abbreviation for newton-metre is N m (and not mN, which would mean millinewton—one-thousandth of a newton). The newton-metre is itself given a special name: the *joule* (after James Prescott Joule, the scientist who did a great deal of work on energy and energy conversion); the symbol for joule is J.

Example 4.1

Calculate the work done when a force of 100 N moves through a distance of 5 m along its line of action.
Solution

$$\text{Work done} = \text{force} \times \text{distance moved in direction of force}$$
$$= 100 \times 5 \text{ N m}$$
$$= 500 \text{ J}$$

Note that the distance moved is in the direction of action of the force.

Example 4.2

Calculate the work done in lifting a mass of 10 kg vertically upwards through a height of 7 m.
Solution   To lift any mass vertically upwards it is necessary to exert a force *equal and opposite* to the weight of the mass. The weight of any body is the force of gravity exerted on it by the Earth. In SI units force is equal to mass × acceleration, and since at any one place the acceleration due to gravity is constant (of the order of 9.81 m/s²) we obtain the weight of any body by multiplying its mass by a constant equal to the acceleration due to gravity at the place at which the weight is being measured.

Taking the acceleration due to gravity (symbol g) as 9.81 m/s², then

$$\text{weight of 10 kg mass} = 10 \times 9.81$$
$$= 98.1 \text{ N}$$

Thus a force of 98.1 N must be exerted vertically upwards on the 10 kg mass to lift it. This force is exerted on the body as it moves vertically upwards (in the direction of the force) a distance of 7 m. Thus the work done in lifting the 10 kg mass through 7 m is given by

$$98.1 \times 7 \text{ newton-metres} = 686.7 \text{ J}$$

Example 4.3

A crane lifts a mass of 500 kg vertically upwards through a distance of 8 m. The rope to which the mass is attached has a mass of 1 kg per metre length and is of length 15 m between the point of attachment to the mass and the drum situated vertically above the mass on to which the rope is wound. Neglecting opposition to motion other than that due to the weight of mass and rope, calculate the work done.
Solution   The weight of the mass to be lifted is the weight of the 500 kg body plus the weight of the rope, initially of length 15 m. The rope length is changing and so therefore is the mass of rope and the weight of rope. The required force, which is equal and opposite to the combined weight, will also be changing.

When the rope length is 15 m

$$\text{rope mass} = 15 \times 1 \text{ kg}$$
$$\text{rope weight} = 15g \text{ N}$$
$$\text{body weight} = 500g \text{ N}$$
$$\text{total weight} = 515g \text{ N}$$

When the rope length is 14 m

rope mass	$= 14$ kg
rope weight	$= 14g$ N
body weight	$= 500g$ N
total weight	$= 514g$ N

Similarly, when the rope length is 13 m the total weight is $513g$ N, when the rope length is 12 m the total weight is $512g$ N, and so on. The total weight, and therefore the total force required, drops by g newtons for every metre moved by the body. The values of force and distance moved by force are as follows.

Force (N $\times g$)	Distance (m)
515	0
514	1
513	2
512	3
511	4
510	5
509	6
508	7
507	8

The graph is shown in figure 4.1. The total work done is the area under the graph between distance = 0 and distance = 8 m, determined by using the figures shown (not actual lengths). The area is that of a trapezium and is equal to

$\frac{1}{2} \times$ sum of parallel sides $\times$ distance between them
$= \frac{1}{2} \times (515 + 507) \times 8 \times 9.81$

(taking g as 9.81 m/s^2) which is 40 496 J or 40.496 kJ.

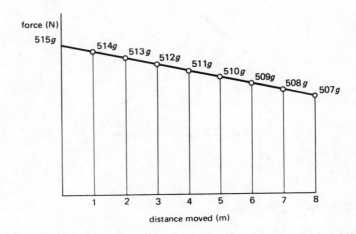

Figure 4.1

Work Done by a Torque

When a force is applied so that a turning effect is obtained, we say that a *torque* is being applied. Many examples of torque occur in engineering and in everyday life. When a door is opened by pushing it on the side opposite the hinges, a torque is being applied to the door; tightening a nut with a spanner exerts a torque on the nut, and there are many other examples.

Torque is measured by the product of the force exerted and the distance between the point of application of the force and the turning centre, along a line at right-angles to the line of application of the force. In figure 4.2, a force F is applied at a distance r from a centre O; this figure could represent a spanner length r tightening a nut at O, the force F being exerted at the spanner end opposite to O. The torque here is $F \times r$ newton-metres, provided that F is measured in newtons and r in metres.

We can see that since the total turning effect, or torque, is the product of the two quantities force and distance, to obtain a given torque a greater value of r means a smaller value of F. This is a useful thing to remember practically, one example being the untightening of a difficult nut when only a limited force can be

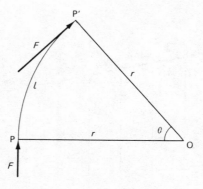

Figure 4.2

applied (determined by the strength of the person doing the work!). If the length of the spanner is increased, for example, by slipping a suitable piece of tubing over the end, the same force applied will now undo the tight nut because the torque has been increased because of the additional distance between the turning centre and the point of application of the force.

Since the force moves its point of application when a torque is applied, work is done. The work done is equal to the product of the force and the distance moved by the point of application of the force in the *direction of action* of the force. Returning to figure 4.2, if the arm length r is moved from position OP to position OP′ then the point of application of the force is moved from point P to point P′ along a distance l. The work done is then $F \times l$ newton-metres. Now, if the angle θ through which the arm moves is measured in radians then the length of arc l is equal to $r \times \theta$.

Thus the work done is given by Fl which equals $Fr\theta$, but the torque applied is Fr, so we can say that the work done is given by the product torque × angle turned through. For this to be so all units must be consistent: F in newtons, r in metres and θ in radians.

Example 4.4

Calculate the work done when a force of 250 N is applied to a spanner of length 10 cm in the normal way and the spanner turns through an angle of 180°.
Solution

$$\text{Torque applied} = \text{force} \times \text{radius}$$
$$= 250 \times 0.1$$

(spanner length in metres!)

$$= 25 \text{ N m}$$

Angle turned through is 180°; this must be converted to radians

$$180° = \pi \text{ radians}$$

so that

$$\text{work done} = \text{torque} \times \text{angle}$$
$$= 25 \times \pi \text{ N m}$$
$$= 78.57 \text{ J}$$

Example 4.5

A pulley being used as part of a lifting machine has a diameter of 25 cm. The force applied tangentially at the rim of the pulley is 50 N. Calculate the work done in 30 revolutions of the pulley.
Solution

$$\text{Torque applied} = \text{force} \times \text{radius}$$
$$= 50 \times 0.125$$
(radius in metres)
$$= 6.25 \text{ N m}$$

For each revolution the angle turned through is 2π radians; for 30 revolutions the angle turned through is $30 \times 2\pi$ radians. Therefore

$$\text{work done} = \text{torque} \times \text{angle turned through}$$
$$= 6.25 \times 30 \times 2\pi \text{ N m}$$
$$= 1178.6 \text{ J}$$

Note that although the units of torque and work are *dimensionally* the same (that is, they are both newton-metres) they are *not* the same quantity. The unit joule is used for work but may not be used for torque for this reason; to express torque in work units it must first be multiplied by an angle in radians, as we have seen.

ENERGY

Energy is the ability to do work. A comprehensive definition is 'the ability to set up a force and in so doing to change, convert or modify a body's physical state, shape or mass or state of rest or of uniform motion in a straight line'. This definition in fact includes the definition of a force (which changes a body's state of rest or of uniform motion in a straight line). There are many kinds of energy: mechanical, electrical, chemical, heat, light, to name but a few. Basic Newtonian physics (based on the work of Isaac Newton) indicates that the amount of energy in the universe is constant and that energy cannot be created or destroyed, only converted from one form to another. However, the work of Albert Einstein earlier in the twentieth century has shown that mass and energy are convertible from one to the other (atomic energy). Newton's basic ideas are sufficiently accurate for the majority of calculations in engineering science and we may neglect any mass–energy conversion (or vice versa) and assume that the energy present in any situation is converted from one form to another as stated by Newton. The unit of energy, not surprisingly, is the same as that of work: the newton-metre or joule. All energy, be it electrical, mechanical or chemical, is measured in the same units when the SI system is used; this is not the case with many other systems of units.

POWER

Power is the rate of doing work, that is, the rate of using or converting energy. The units are energy units per unit time. In the SI System the unit of power is the joule per second (newton-metre per second), which is given the special name the *watt*, symbol W. (This unit is named after James Watt, the Scottish engineer who did a great deal of work on energy and energy conversion, notably on steam engines.) As with other units we also have multiples: the kilowatt (kW) and the megawatt (MW), and sub-multiples: the milliwatt (mW), microwatt (μW), and so on.

Power is calculated by determining the rate of change of energy with time. Average power may be determined by dividing the energy converted by the time taken, and instantaneous power may be determined by graphical means (the slope of the energy–time curve) or by mathematical means, using the integral calculus. All power calculations use the same unit of power whatever the nature of the energy involved. A unit once commonly used and still quoted in some mechanical uses—the horsepower—belongs to the Imperial Gravitational System and should not be used in conjunction with SI units; it can be shown that 1 horsepower is equivalent to 746 watts.

Example 4.6

Calculate the output power of a crane motor which raises a body of mass 1 tonne to a height of 18 m in 15 s. Take $g = 9.81$ m/s^2.
Solution

$$\text{Mass of body} = 1000 \text{ kg}$$
$$\text{Weight of body} = 1000 \times 9.81$$
$$= 9810 \text{ N}$$

$$\text{Force required to lift body} = 9810 \text{ N}$$

$$\begin{matrix}\text{Work done in raising} \\ \text{body through a height of 18 m}\end{matrix} = 9810 \times 18$$
$$= 176\,580 \text{ J}$$

The work is done in 15 s, so the average rate of working, or average power, is

$$\frac{176\,580}{15} \text{ J/s} = 11\,772 \text{ W}$$

which equals 11.772 kW.

Useful Relationships Involving Power

As we have seen, work is determined by multiplying force by distance

work = force × distance

the distance being that moved by the point of application of the force in the direction of the line of action of the force. Thus the rate of working, that is, the rate of change of work or energy (with time) is given by

rate of change of work = rate of change of (force × distance)

and for a constant force

rate of change of work = force × rate of change of distance
= force × velocity

But the rate of change of work or energy is the power, so that

power = force × velocity

Similarly, in rotational movement, we saw that the work done by a torque is given by

work done by torque = torque × angle

Therefore rate of change of work done by a torque is given by

rate of change of work = rate of change of torque × angle

and for a constant torque

rate of change of work = torque × rate of change of angle

so that

power = torque × angular velocity

which is a similar relationship to the one above for linear velocity.

These relationships are useful in problems where velocities are known. Note that they apply to instantaneous or average values of the quantities concerned.

Example 4.7

A train of mass 300 tonnes travels along a level track at 100 km/h. The resistive force opposing motion is 150 N/tonne. Determine the power required from the engine.
Solution

$$\text{Velocity of train} = 100 \text{ km/h}$$
$$= \frac{100\,000}{3600} \text{ m/s}$$
$$= 27.78 \text{ m/s}$$

The only force required is that to overcome the total resistive force, which equals 300×150, that is, 45 000 N. Thus

$$\text{power required} = \text{force} \times \text{velocity}$$
$$= 45\,000 \times 27.78$$
$$= 1\,250\,100 \text{ W}$$
$$= 1.25 \text{ MW}$$

EFFICIENCY

Whenever energy conversion takes place in a machine or some other electrical or mechanical device, there are nearly always losses of one kind or another. Consider, for example, an electric motor, which converts electrical energy (from an electricity supply) to mechanical energy. The conventional motor consists of a shaft which rotates; the shaft must be supported by bearings which allow rotation with minimum resistance to motion and, when turning, the air within the motor must be pushed out of the way as the shaft turns. There are two sets of losses here: one due to *friction* at the bearings—which however small still exists—and one due to *windage*—the loss due to resistive forces set up by the air. Again, design of the rotating shaft (rotor) can reduce this loss but it will always be present to some extent. Some of the input energy, then, will be lost because of these losses and will therefore not be available at the output of the motor. (Other losses occur within the machine but these will be ignored for the moment.)

The efficiency of an energy-conversion system is a measure of the amount of energy available at the output compared to the energy put in at the input, that is

$$\text{efficiency} = \frac{\text{work or energy output}}{\text{work or energy input}}$$

Since the numerator and denominator of the right-hand side of the equation have the same units, the efficiency has no units but is a pure ratio. As such its value will always be less than unity, since the output is always smaller than the input. Expressed in the above manner the efficiency is called the *per unit* efficiency, since it tells us the fraction of each input unit of energy that is obtained at the output. If the ratio is multiplied by 100 the efficiency is then being given as a percentage.

Another way of writing efficiency, which is particularly useful when the output is very nearly equal to the input (that is, the efficiency is very high) is as follows.

$$\text{Efficiency} = \frac{\text{output energy}}{\text{input energy}}$$

but since the input energy is equal to the output energy plus the losses within the system

$$\text{efficiency} = \frac{\text{input energy} - \text{losses}}{\text{input energy}}$$

$$= 1 - \frac{\text{losses}}{\text{input energy}}$$

In practice, efficiencies range from very low figures to almost 100 per cent. The transformer, which converts electrical energy at one level of voltage and current to another level of voltage and current and has no moving parts, is particularly efficient (97 to 99 per cent), while the earlier forms of steam engine had efficiencies as low as 10 per cent. This last figure tells us that only one-tenth of the input energy appears at the output, the remaining nine-tenths being converted to some other form of non-useful energy and thus, to all intents and purposes, being lost.

Example 4.8

An electric crane raises a body of mass 1 tonne through 4 m in 20 s; the motor driving the crane is supplied with 2.5 kW. Determine the efficiency of the system ($g = 9.81$ m/s^2).
Solution

$$\begin{aligned}\text{Work done in lifting a body} &= \text{weight of body} \\ &\quad \times \text{height lifted} \\ &= 1000 \times 9.81 \times 4 \\ &= 3924 \text{ J}\end{aligned}$$

The motor is provided with 2500 W, that is, 2500 joules per second, for 20 s to do this work. Therefore

$$\begin{aligned}\text{energy input} &= 2500 \times 20 \\ &= 5000 \text{ J}\end{aligned}$$

Therefore efficiency, which is output energy/input energy, is given by

$$\begin{aligned}\text{efficiency} &= \frac{3924}{5000} \\ &= 0.7848 \text{ per unit} \\ &= 78.48 \text{ \%}\end{aligned}$$

ASSESSMENT EXERCISES

Long Answer

(Take $g = 9.81$ m/s²)

4.1 Calculate the work done in lifting a mass of 50 kg vertically upwards through a height of 20 m.

4.2 The work done in raising a certain mass through a height of 10 m is 750 J. Determine the mass in kg.

4.3 Determine the work done in accelerating a vehicle of mass 1.2 t with constant acceleration of 1.2 m/s² for a distance of 45 km along a level road. Neglect any opposition to motion.

4.4 A crane lifts a mass of 0.5 t vertically upwards through a distance of 10 m. The rope attached to the mass is of length 18 m between the mass and the crane drum on to which the rope is wound. The total work done is 58 kJ. Calculate the rope mass per metre length.

4.5 The work done in turning a spanner of length 12 cm through 90° is 60 J. Determine the value of the applied force.

4.6 Calculate the output power of a winch used to raise a body of mass 1.5 t through a vertical height of 25 m in 20 s.

4.7 If the vehicle in question 4.3 moves the distance given in a time of 30 min along the level, find the output power of the vehicle motor in each case.

4.8 A crane supplied with 3 kW of power raises a body of mass 450 kg through 10 m. The crane efficiency is 78 per cent. Calculate the time taken for the body to be lifted.

4.9 An electric crane raises a body of mass 2 t through 7.5 m in 34 s. Calculate the input power if the efficiency of the system is 82 per cent. ($g = 9.81$ m/s².)

4.10 Determine the power dissipated when a force of 150 N acting at an angle of 30° to the horizontal pulls a mass 15 m along a horizontal surface in a time of 8 s.

Short Answer

4.11 Calculate the work done when a 1-kN force moves a distance of 5 m along its line of action.

4.12 A force moves its point of application 3.2 m along its line of action and the work done is 1500 J. Calculate the value of the force.

4.13 12 kJ of work are done by a force of 1000 N in moving a certain distance along its line of action. Calculate the distance moved.

4.14 Calculate the power used if 3.4 kJ of work are done in 1.3 s.

4.15 Calculate the time taken to do 1500 kJ of work if the power involved is 150 kW.

4.16 Calculate the work done by a machine working at 4.7 kW for 3.7 min.

4.17 Calculate the efficiency of a machine taking 25 kW at the input and yielding 23.7 kW at the output.

4.18 The per-unit efficiency of a machine is 0.72. Calculate the input power for a 12 kW output.

4.19 If the input power to the machine in question 4.18 is 10 kN, calculate the output power.

4.20 Calculate the work done by a force which changes linearly from 1000 N to 5000 N and moves 100 m while the change is taking place. Assume the movement to be along the line of action of the force.

Multiple Choice

4.21 The work done when a force of 500 N moves a distance of 4 m along its line of application is equal to
   A. 19.62 kJ   B. 125 W   C. 2 kJ   D. 2 kW

4.22 The work done in lifting a 10 kg mass vertically upwards through a height of 7 m is equal to
   A. 70 J   B. 686.7 J   C. 1.43 J   D. 14.01 J

4.23 A force moves a distance of 5 m along its line of action and 2 kJ of work are done. The force is equal to
   A. 2.5 kN   B. 2.5 N   C. 400 N   D. 10 kN

4.24 A force of 1.1 kN moves through a distance of 0.5 m in 4 s. Assuming the movement is along the line of action of the force the power used is
   A. 8800 W   B. 2200 W   C. 137.5 J   D. 137.5 W

4.25 A body of weight 100 N is lifted through 1.5 m in 0.4 s. The power required is
   A. 38.23 W   B. 26.67   C. 375 W   D. 60 W

4.26 A linearly increasing force acts over a distance of 15 m in the direction of action of the force. The force is of magnitude 5 N initially and has increased to 55 N by the time the point of application has moved 15 m. The work done is equal to
   A. 75 J   B. 450 J   C. 900 J   D. 825 J

4.27 If the time taken for the force described in question 4.26 to move the full distance is 10 s, the power dissipated is
   A. 45 W   B. 4.5 kW   C. 90 W   D. 82.5 W

4.28 The power input to a certain machine is 1.1 kW, the power output is 0.9 kW. The per-unit efficiency is
   A. 82   B. 0.82   C. 0.2   D. 20

4.29 The efficiency of a mechanical system is 75 per cent. The input power is 1500 J/s. The output power is

   A. 2000 J/s   B. 1.125 kW   C. 112.5 W   D. 2 kW

4.30 The output power of a certain machine is 15 kW. The per-unit efficiency is 0.9. The input power is
   A. 166.67 W   B. 1350 kW   C. 13.5 kW   D. 16.6 kW

5 Heat

Part of the definition of energy is that it is the ability to change,
convert or modify a body's state, shape or mass or state of rest.
Heat is one form of energy and as such is measured in energy
units: joules, the symbol for which is J.

Heat energy can be transferred from place to place by one or
more of various methods considered in more detail below, and may
be converted to other energy forms, including electrical, mechanical and chemical. Heat energy may be derived from other energy
forms, as in the electric fire, the incandescent lamp, friction, impact
of moving bodies, chemical reactions, and so on.

HEAT TRANSFER

Heat energy may be transferred from one place to another in any of
three ways. These are *conduction*, which takes place mainly in
solids and liquids, *convection*, which occurs mainly in gases and
liquids and *radiation*, which requires no physical medium but can
occur in solids, liquids and gases.

Conduction

All matter may be regarded as being made up of millions of tiny
particles (molecules) which are constantly moving. These particles
thus have energy of movement: kinetic energy. If a bar of suitable
material is placed with one end in a heat source, such as a fire or
flame, it is found that, after a time, the end remote from the heat
source gets warm. The energy of the particles of matter at the heat
source end is increased and some of this energy is passed to
adjacent particles, until eventually the energy content as a whole
rises, that is, the whole body gets warmer. No particles actually
move down the bar although we sometimes use the expression
'heat flow' to describe the transfer of energy from one part to the
other. This method of heat transfer is called *conduction*. It occurs
mainly in matter which is made up of particles in close proximity,
such as solids and liquids.

Convection

When heat energy transfer takes place because of actual movement of the particles containing the heat energy the process is known as *convection*. The best-known example is that of warm air, which moves from one place to another (usually rising) taking the heat energy with it. A movement of heated particles in the manner described is called a convection current; convection currents may be observed when heating a liquid by adding a coloured dye to the liquid.

Radiation

Heat energy may be *radiated* by means of *electromagnetic waves*. Whenever electric and magnetic fields act together such that one is at right-angles to the other, energy is transferred. The nature of the energy depends on the characteristics of the electric and magnetic fields and may be heat, light, radio, ultraviolet, infrared, and so on. The energy we receive from the Sun, which is mainly light and heat energy, is conveyed in this manner, called *radiation*.

Radiation requires no physical medium but can pass through a number of different materials, depending on their atomic construction. Certain surfaces more easily absorb radiated energy while others reflect it. Black surfaces absorb and, if they are on a hot body, radiate more effectively than others; white or silvered surfaces reflect radiated energy. The reflector on an electric fire is silvered so that the radiated heat energy from the fire bar is emitted from the fire front. (It is interesting to note that so-called radiators in central-heating systems do not, in fact, radiate but transfer their heat energy by conduction and convection through the surrounding air.)

TEMPERATURE

Temperature and heat are not the same thing, although the terms are often loosely used to convey the same idea. The heat of a body is a measure of its total energy, whereas temperature ('the degree of hotness') is a measure of the *average* energy of the particles making up the body.

There are several *scales* of temperature available. A scale of temperature is a means by which temperatures may be compared with each another. Probably the best-known temperature scales are the Fahrenheit scale and the Centigrade scale, the latter being more correctly called the Celsius scale after the man who originated it. In devising the early temperature scales, much use was made of various temperature points thought to be fixed, the main ones being the boiling point of water and the melting point of ice. In fact, the temperatures at which these events occur depend on external physical conditions (the pressure of the surrounding air, for example) and are not suitable points for accurate definition. The Fahrenheit scale and the Celsius scale both took the melting point of ice and the boiling point of water as their fixed points, the range between the two being divided into a number of equal intervals called 'degrees'. The Fahrenheit scale has 180 degrees between the two points, the melting point of ice being taken as 32 degrees, written $32\,°F$, and the boiling point of water being taken as $212\,°F$. The Celsius scale puts the lower temperature point as zero, written $0\,°C$, and the upper as $100\,°C$. Both scales are still in common use, particularly the Celsius scale (although it is usually referred to as Centigrade) since, in an extended form, it is the SI temperature scale.

The temperature scale in the SI System of units is the *Kelvin* scale. The fixed points on this scale are not those of the other scales but are much more accurately reproducible. They are *absolute zero* and the *triple point of water*. Absolute zero is the point at which an ideal gas is at zero pressure (an ideal gas being one in which the molecules are so small that their size can be ignored). The triple point of water is the point at which water, water vapour and ice exist in stable equilibrium. Between these points the temperature range is divided up into 273.16 equal intervals called *kelvins* (not degrees kelvin); absolute zero occurs at zero kelvin, written $0\,K$, and the triple point occurring at $273.16\,K$. At normal temperature and pressure the melting point of ice occurs at $273.15\,K$ and the boiling point of water occurs at $372.15\,K$, so there are 100 kelvins between the melting point of ice and the boiling point of water. Clearly, then, one kelvin temperature interval is equal to one degree in the Celsius scale, so the Celsius scale can be used instead of the Kelvin scale, the difference being that temperature in Celsius is equal to temperature in kelvin *less* 273.15. Absolute zero occurs at $-273.15\,°C$ and the triple point at $0.01\,°C$.

SPECIFIC HEAT CAPACITY

The *specific heat capacity* of a material is the heat energy required to change the temperature of one unit of mass of the material by one degree or one temperature interval. The symbol is *s*. The unit is either kilojoule per kilogram per degree Celsius or kilojoule per kilogram per kelvin (the numerical value being the same in either units since one Kelvin temperature interval is equal in size to one Celsius degree). The abbreviations used for these units are $kJ/kg\,°C$ or $kJ/kg\,K$.

The relative specific heat capacity of a material is its specific heat capacity divided by the specific heat capacity of water. Alternatively

specific heat capacity = relative specific heat capacity
× specific heat capacity of water

The specific heat capacity of water (to one decimal place) is $4.2\,kJ/kg\,°C$. Relative specific heat capacity was formerly known only as 'specific heat' and this term is still occasionally encountered; its use is not recommended.

Heat-energy Calculation

The specific heat capacity of a substance is the heat energy which changes the temperature of unit mass by one degree. Thus the heat energy involved when a mass of *m* units has its temperature changed from t_1 degrees to t_2 degrees is given by

$$\text{heat energy} = s \times m \times (t_2 - t_1)$$

where *s* is the specific heat capacity. In the SI System the unit of *m* is the kilogram, of t_1 and t_2 kelvin or degree Celsius (if kelvin then the word 'degree' above should strictly be written 'temperature interval') and of *s* is the $kJ/kg\,°C$ or $kJ/kg\,K$.

Example 5.1

Calculate the heat required to change the temperature of 5 kg of water from $10\,°C$ to $60\,°C$. The specific heat capacity of water is $4.2\,kJ/kg\,°C$.
Solution

$$\text{Heat energy} = 4.2 \times 5 \times (60 - 10)$$
$$= 1050\,kJ$$

Example 5.2

Calculate the heat energy required to change the temperature of a block of metal of mass 25 kg from 350 K to 420 K. The relative specific heat capacity of the metal is 0.2.
Solution

$$\text{Specific heat capacity of metal} = 0.2 \times 4.2$$
$$= 0.84\,kJ/kg\,K$$

(taking specific heat capacity of water as $4.2\,kJ/kg\,K$), thus

$$\text{heat energy required} = 0.84 \times 25 \times (420 - 350)$$
$$= 0.84 \times 25 \times 70$$
$$= 1470\,kJ$$

The same method of calculation is used whether the heat energy is being given *to* a body to raise its temperature or whether the heat energy is being given out *from* the body when its temperature is falling. For the same temperature interval the same heat energy is involved.

THERMAL CAPACITY AND WATER EQUIVALENT

The amount of heat energy required to change the temperature of any particular body by one degree or one temperature interval is given by

$$\text{heat required per unit temperature} = \text{mass} \times \text{specific heat capacity}$$

since the specific heat capacity is the heat required per unit mass for a one degree temperature change. This value of heat energy for a particular body is called the *thermal capacity* of the body. For a body made up of various materials the thermal capacity is the sum of the thermal capacities of the individual masses making up the body. The SI unit is kJ/°C or kJ/K.

The *water equivalent* of a body is the mass of water which would experience the same change in temperature as the body for a given amount of heat energy. We saw earlier that for a particular substance

$$\text{relative specific heat capacity} = \frac{\text{specific heat capacity of substance}}{\text{specific heat capacity of water}}$$

Now, the specific heat capacity of a substance is the amount of heat involved in unit temperature change per unit mass, so that, as we have seen, for a particular temperature change and a particular mass, the heat involved is

mass × temperature change × specific heat capacity

For the body

heat energy = specific heat capacity of body
× mass of body × temperature change

For water

heat energy = specific heat capacity of water
× mass of water × temperature change

If the heat energy is the same in each case : specific heat capacity of body × mass of body × temperature change *is equal to* specific heat capacity of water × mass of water × temperature change and for the same temperature change

$$\text{specific heat capacity of body} \times \text{mass of body} = \text{specific heat capacity of water} \times \text{mass of water}$$

so

$$\text{mass of water} \times \text{mass of body} = \frac{\text{specific heat capacity of body}}{\text{specific heat capacity of water}}$$

$$= \text{relative specific heat capacity} \times \text{mass of body}$$

The mass of water in this case is the water equivalent as defined above so that

$$\text{water equivalent} = \text{mass of body} \times \text{relative specific heat capacity}$$

As an example, suppose the relative specific heat capacity of a certain material were 0.5, then the heat required per unit mass per unit temperature change for water is *twice* that required per unit mass per unit temperature change for the material. It follows then that to obtain the *same* temperature change in water for a given amount of heat energy the mass of water must be only half that of the mass of the body.

Example 5.3

Determine the thermal capacity and water equivalent of 5 kg of copper if its relative specific heat capacity is 0.09 and the specific heat capacity of water is 4.2 kJ/kg °C.
Solution

Thermal capacity = mass × specific heat capacity of copper
= mass × relative specific heat capacity of
copper × specific heat capacity of water
= 5 × 0.09 × 4.2
= 1.89 kJ/°C

Water equivalent = mass of copper × $\dfrac{\text{relative specific}}{\text{heat capacity}}$

$$= 5 \times 0.09$$
$$= 0.45 \text{ kg}$$

Example 5.4

A body of mass 25 kg consists of two equal parts, one of copper and one of steel. The relative specific heat capacities for copper and steel are 0.09 and 0.12 respectively. Calculate the thermal capacity of the body and the water equivalent of each part of the body.

Solution   With compound bodies like this, calculations are made separately for each component part and then appropriate additions made.

For copper

specific heat capacity = 0.09×4.2 kJ/kg K

(taking specific heat capacity of water as 4.2 kJ/kg K)

$$= 0.378$$
thermal capacity = 12.5×0.378

(mass is $\frac{1}{2} \times 25$ kg)

$$= 4.725 \text{ kJ/K}$$

For steel

specific heat capacity = $0.12 \times 4.2 = 0.504$
thermal capacity = 0.504×12.5
$$= 6.3 \text{ kJ/K}$$

Thermal capacity of body is therefore $4.725 + 6.3$ kJ/k, that is, 11.025 kJ/K.

Water equivalent of copper = 0.09×12.5
$$= 1.125 \text{ kg}$$

Water equivalent of steel = 0.12×12.5
$$= 1.5 \text{ kg}$$

The following examples illustrate the method of approach used in calculations and problems involving mixtures. In most of these problems involving bodies (or parts of bodies) losing heat and other bodies gaining heat, the assumption is usually made that the heat lost by the one equals the heat gained by the other, unless, of course, the amount of heat lost to the surroundings is known and can be taken into consideration.

Example 5.5

A copper container of mass 0.1 kg holds 0.05 kg of water at a temperature of 20 °C. A piece of metal of mass 0.04 kg at a temperature of 90 °C is immersed in the water and the water temperature rises to 25 °C. The relative specific heat capacity of copper is 0.1 and the specific heat capacity of water is 4.2 kJ/kg °C. Assuming negligible heat loss to the surroundings, calculate the relative specific heat capacity of the metal which was added to the water.

Solution

$\begin{array}{l}\text{Heat energy transferred} \\ \text{from metal}\end{array} = \begin{array}{l}\text{specific heat capacity} \times \text{mass} \\ \times \text{temperature change}\end{array}$
$$= s_r \times 4.2 \times 0.04 \times (90 - 25)$$

where s_r is the required relative specific heat capacity. Note that the final temperature of the metal is that of the water, 25 °C.

$$= 10.92 s_r \text{ kJ}$$

heat energy gained by water = mass × specific heat capacity
$$\times \text{ temperature change}$$
$$= 0.05 \times 4.2 \times (25 - 20)$$
$$= 1.05 \text{ kJ}$$

$$\begin{aligned}\text{heat energy gained}\\ \text{by copper container}\end{aligned} = \begin{aligned}&\text{mass} \times \text{specific heat capacity}\\ &\times \text{temperature change}\end{aligned}$$

$$= 0.1 \times 0.1 \times 4.2 \times (25 - 20)$$
$$= 0.21 \text{ kJ}$$

Note that the copper container experiences the same temperature change as the water.

Total heat gained $= 1.05 + 0.21 = 1.26$ kJ
heat lost by metal $= 10.92 s_r$ kJ

Assuming negligible loss to the surroundings

heat lost by metal = heat gained by water and container
$$10.92 s_r = 1.26$$

hence

$$s_r = \frac{1.26}{10.92}$$
$$= 0.11$$

The relative specific heat capacity of the metal in this example is 0.11.

The method illustrated in this example is a standard method for calculating specific heat capacity and relative specific heat capacity. The copper container mentioned is called a *calorimeter* in this application, the field of heat energy measurement using the method being called *calorimetry*.

Note that throughout the calculation use was made of the relationship

$$\begin{aligned}\text{specific heat capacity}\\ \text{of a material}\end{aligned} = \begin{aligned}&\text{relative specific heat capacity}\\ &\times \text{specific heat capacity of water}\end{aligned}$$

Example 5.6

A metal bar of mass 200 kg at a temperature of 500 °C is immersed in a tank containing 1.5 tonnes of oil at a temperature of 20 °C. The final temperature of the oil and bar is 38.5 °C. The relative specific heat capacity of the metal is 0.15. Neglecting heat losses to the tank and surrounding atmosphere calculate the relative specific heat capacity of the oil.

Solution   Let the relative specific heat capacity of the oil be s_r. Then

$$\begin{aligned}\text{heat gained by oil} &= \text{mass} \times \text{specific heat capacity}\\ &\quad \times \text{change in temperature}\end{aligned}$$
$$= 1500 \times s_r \times 4.2 \times (38.5 - 20)$$
$$= 116\,550 s_r \text{ kJ}$$
$$\text{heat lost by metal} = 200 \times 0.15 \times 4.2 \times (500 - 38.5)$$
$$= 58\,149 \text{ kJ}$$

Neglecting losses, heat lost = heat gained, therefore

$$116\,550 s_r = 58\,149$$
$$s_r = \frac{58\,149}{116\,550}$$
$$= 0.498$$

Note that the value 4.2 taken as the specific heat capacity of water is not required, since it appears on both sides of the equation.

Example 5.7

In an experiment involving the immersion of a heated body in a calorimeter containing oil the following observations were made

mass of body 0.016 kg
relative specific heat capacity of body 0.12
temperature prior to immersion 60 °C
mass of oil 0.030 kg
relative specific heat capacity of oil 0.4
initial temperature of oil 17 °C
final temperature of oil and body 19.5 °C
heat lost to surroundings 100 J

Determine the water equivalent of the calorimeter assuming the specific heat capacity of water to be 4.18 kJ/kg °C.

Solution The heat lost by the body is transferred to the oil, to the calorimeter and to the surroundings. Assuming no other losses we can equate the heat lost by the body to the sum of the heat energies gained by these three.

Let the water equivalent of the calorimeter be W kg. Since the water equivalent is relative specific heat capacity × mass, and since the heat energy involved when a body changes temperature is given by

$$\begin{aligned} \text{heat energy} &= \text{mass} \times \text{relative specific heat capacity} \\ &\quad \times \text{specific heat capacity of water} \\ &\quad \times \text{change in temperature} \\ &= \text{water equivalent} \times \text{specific heat capacity of} \\ &\quad \text{water} \times \text{change in temperature} \end{aligned}$$

so that

$$\text{heat gained by calorimeter} = W \times 4.18 \times (19.5 - 17)$$

(oil and calorimeter initial temperature the same)

$$= 10.45\,W\,\text{kJ}$$

$$\begin{aligned} \text{heat gained by oil} &= 0.030 \times 0.4 \times 4.18 \times (19.5 - 17) \\ &= 0.1254 \text{ kJ} \end{aligned}$$

heat lost to surroundings = 0.1 kJ

$$\begin{aligned} \text{total heat transferred to} \\ \text{surroundings, oil and calorimeter} \end{aligned} \;\begin{aligned} &= 0.1 + 0.1254 + 10.45\,W \\ &= 0.016 \times 0.12 \\ &\quad \times 4.18(60 - 19.5) \\ &= 0.325 \text{ J} \end{aligned}$$

Thus

$$0.325 = 0.1 + 0.1254 + 10.45\ W$$

and

$$10.45\,W = 0.0996$$
$$W = 0.0095 \text{ kg}$$

The water equivalent of the calorimeter is 0.0095 kg (9.5 grams).

LATENT HEAT (SPECIFIC ENTHALPY)

When a substance undergoes a change of state from solid to liquid or liquid to gas, a specific amount of heat energy is required which is absorbed by the material *without change of temperature*. Similarly as the substance cools and the state changes from gas to liquid or liquid to solid, the heat energy absorbed is then given out, again without change of temperature. This additional heat to that required for change in temperature is called *latent* (hidden) heat.* Another name is specific enthalpy ('enthalpy' is a measure of the energy of a body or system which is contained within the body or system). Latent heat is usually quoted in energy units per unit mass. The type of latent heat is also given, either *latent heat of evaporation*—taken *in* when a body changes from liquid to gas and given *out* when the process is reversed—and *latent heat of fusion*—taken *in* when the body changes from solid to liquid and given *out* when the process is reversed. It must be emphasised that the taking in or giving out of latent heat occurs at constant temperature, either at melting point (fusion) or boiling point (evaporation).

Considering water as an example : if ice is heated it will absorb about 2.1 kJ/kg for each degree rise in temperature until 0 °C is reached (the specific heat capacity of ice is approximately half that of water). At melting point each kilogram of ice will absorb approximately 335 kJ while it melts, the temperature remaining at 0°C. If heating continues, each kilogram of water then absorbs 4.2 kJ per degree rise until boiling point is reached at 100°C. Here, each kilogram absorbs about 2000 kJ as it turns into steam, the temperature remaining constant; further heating produces superheated steam. The figures given are said to be 'about' or 'approximately' because actual figures are subject to external conditions

---

* The heat absorbed or given out when the temperature of a body *changes* (either increases or decreases) is called *sensible* heat; the word 'sensible' here is an engineering term and does not have its usual meaning.

(such as pressure). The following examples illustrate how latent heat is taken into consideration in temperature calculations and other heat energy problems.

Example 5.8

Ice of mass 0.1 kg at a temperature of $-10\,°C$ is added to 1 kg of water at $20\,°C$ contained in a vessel with a water equivalent of 0.03 kg. The relative specific heat capacity of ice is 0.5 and the latent heat of fusion may be taken as 335 kJ/kg. The specific heat capacity of water is 4.2 kJ/kg °C. Calculate the final temperature of the water.

Solution   Here, the ice requires sufficient heat to raise its temperature to melting point, the latent heat of fusion taken in while the ice is melting at $0\,°C$, and then further heat to raise the temperature of the resulting water formed from the ice up to the final temperature. All this heat is derived from the water which was initially in the container and from the container itself; in the process of giving out this heat the water and container temperature falls from $20\,°C$ to the final value, which we have to determine.

Let t represent the final temperature in degrees Celsius.

Heat energy lost by water = mass × specific heat capacity
$$\times \text{ change in temperature}$$
$$= 1 \times 4.2 \times (20 - t)$$
$$= 84 - 4.2t \text{ kJ}$$

Heat energy lost by container = water equivalent × specific
$$\text{heat capacity of water}$$
$$\times \text{ change in temperature}$$
$$= 0.03 \times 4.2 \times (20 - t)$$
$$= 2.52 - 0.126t \text{ kJ}$$

(Note that the water equivalent could be added directly to the water mass—this is the mass of water corresponding to the container.)

Heat energy required by
ice in changing temperature = mass × relative
from $-10°C$ to $0°C$ 　　　specific heat capacity
　　　　　　　　　　　　　× specific heat
　　　　　　　　　　　　　capacity of water
　　　　　　　　　　　　　× change in temperature

$$= 0.1 \times 0.5 \times 4.2 \times 10$$
$$= 2.1 \text{ kJ}$$

Heat required by ice
in changing state at $0°C = \dfrac{\text{latent heat}}{\text{of fusion}} \times \text{mass}$
(latent heat)

$$= 335 \times 0.1$$
$$= 33.5 \text{ kJ}$$

Assuming no loss of mass, the ice produces 0.1 kg of additional water, and

heat required by this water
in changing from $0°C$ 　　= mass × specific heat capacity
to final temperature 　　　× change in temperature
$$= 0.1 \times 4.2 \times t$$
$$= 0.42t \text{ kJ}$$

Total heat lost by water
and container 　　$= 84 - 4.2t + 2.52 - 0.126t$

$$= 86.52 - 4.326t \text{ kJ}$$

Total heat gained by ice
and water formed from ice $= 2.1 + 33.5 + 0.42t \text{ kJ}$

$$= 35.6 + 0.42t$$

Hence

$$86.52 - 4.326t = 35.6 + 0.42t$$
$$50.92 = 4.746t$$

and

$$t = 10.73\,°C$$

The final temperature is $10.73\,°C$.

Example 5.9

In an experiment to determine the latent heat of evaporation of water, steam is passed into a calorimeter and the following observations are made

mass of calorimeter 0.06 kg
relative specific heat capacity of
   calorimeter material 0.1
initial mass of water 0.07 kg
initial temperature 20 °C
final mass of water 0.0719 kg
final temperature 35 °C

The specific heat capacity of water is 4.186 kJ/kg °C.

Determine from these observations the latent heat of evaporation of water.

Solution In this experiment the steam passed into the calorimeter condenses into water, releasing latent heat energy as it does so. The water so formed then cools to the final temperature, releasing further heat energy. The total heat energy released increases the temperature of the rest of the water already in the container, and the container itself, from 20 °C to 35 °C.

Let L be the latent heat of evaporation of water (kJ/kg), then

latent heat released by steam = mass of steam $\times L$

Now the mass of steam is the difference between the mass of water initially in the calorimeter and the mass of water finally in the calorimeter, that is, $0.0719 - 0.07$ kg. This equals 0.0019 kg (1.9 grams). So heat released by steam as it condenses at $100\,°C$ $= 0.0019L$ kJ

Heat released by the water from steam as it cools from $100\,°C$ to $35\,°C$ is given by

mass of water that is cooling $\times 4.186 \times (100 - 35)$
$= 0.0019 \times 4.186 \times 65$
$= 0.517$ kJ

So

total heat released $= 0.0019L + 0.517$

This heat changes the temperature of the initial water and the calorimeter from 20 °C to 35 °C and is equal to (mass of water + water equivalent of calorimeter) $\times 4.186 \times (35 - 20)$.

Water equivalent of calorimeter = mass $\times$ relative specific heat capacity

So

heat absorbed by water and calorimeter $= (0.07 + 0.06 \times 0.1) \times 4.186 \times 15$
$= 4.772$ kJ

Hence

$$0.0019L + 0.517 = 4.772$$
$$0.0019L = 4.772 - 0.517$$
$$= 4.255$$
$$L = \frac{4.255}{0.0019}$$
$$= 2240 \text{ kJ/kg}$$

(The figure usually quoted is of the order of 2257 kJ/kg.)

Example 5.10

An electric furnace smelts 85 kg of tin every hour from an initial temperature of 15 °C. Find the heat energy required per hour assuming the melting point of tin to be 235 °C and the latent heat of fusion to be 55 MJ/kg. The specific heat capacity of tin is 235 J/kg °C. Use the answer to determine the required furnace power in kilowatts.

Solution The total heat energy required is equal to the heat required to raise the temperature of 85 kg of tin from 15 °C to 235 °C plus the latent heat necessary to convert 85 kg of solid tin to liquid tin at 235 °C.

$$\text{Heat required to raise temperature} = \text{mass} \times \text{specific heat capacity} \times \text{change in temperature}$$
$$= 85 \times 0.235 \times (235 - 15)$$
$$= 4394.5 \text{ kJ}$$

$$\text{Latent heat required for liquefaction} = \text{mass} \times \text{latent heat of fusion}$$
$$= 85 \times 55$$
$$= 4675 \text{ kJ}$$

$$\text{Total heat energy required} = 4394.5 + 4675$$
$$= 9069.5 \text{ kJ}$$
$$= 9.0695 \text{ MJ}$$

This energy is required every hour, that is, every 3600 seconds, thus

$$\text{rate of using energy} = \frac{9069.5}{3600}$$
$$= 2.52 \text{ kJ/s}$$

One kilojoule per second is called one kilowatt, so

$$\text{power required} = 2.52 \text{ kW}$$

EXPANSION DUE TO HEAT

Many materials, particularly metals, change their shape when heated. In general this is due to each side of the body expanding as the heat energy is absorbed by the body material. If allowed to cool, the body resumes its original shape, provided that the increase in temperature has not been so great as to cause permanent distortion. One of the ways in which this fact is used in engineering is in *shrink fits*, where a piece of material to be joined to another is first heated, fitted over the other then allowed to cool. During cooling the piece shrinks and a tight fit is obtained. One example of the method is fitting steel 'tyres' to locomotive wheels.

Although only relatively small expansions are normally encountered (depending of course on temperature rise) they must be taken into account, and so measurements are made of expansions of different materials per temperature interval change. The expansion is taken into account in, for example, the building of bridges, buildings, railway lines and in machinery in general engineering.

If a straight line is scribed on a suitable material subject to expansion when heated, and if the temperature of the material is increased, then the length of the line will increase. If the increase in length is divided by the original length, we have the expansion per unit length. Dividing the expansion per unit length by the change in temperature yields the expansion per unit length per temperature interval change. This quantity is called the *coefficient of linear expansion* for the material. The symbol is usually taken as α (alpha) and the SI unit is usually quoted as millimetre per millimetre per degree Celsius (or per kelvin), abbreviated mm/mm °C. Provided that the expansion and the original length are measured in the same units, the coefficient of linear expansion may also be quoted as a ratio per degree Celsius or per kelvin. The value of the expanded length is given by

$$\text{expanded length} = \text{original length} + (\text{original length} \times \text{coefficient of linear expansion} \times \text{change in temperature})$$

Example 5 11

A metal bar of length 3 m at 15 °C is heated to 80 °C. The coefficient of linear expansion of the metal is 0.000 011/°C. Calculate the length of the bar at the increased temperature.

Solution

New length = original length + expansion due to change in temperature

$$= 3 + 3 \times 0.000\,011 \times (80 - 15)$$
$$= 3 + 0.002\,145$$
$$= 3.002\,145 \text{ m}$$

(An extension of 2.145 mm.)

Consider the area of a rectangle of side lengths l_1 and l_2 metres. It is equal to $l_1 l_2$ square metres. If the surface is heated and undergoes a temperature change of t degrees Celsius or t kelvins, side l_1 will have a new length $l_1 + l_1 \alpha t$, that is, $l_1(1 + \alpha t)$ and, similarly, side l_2 will have a new length $l_2(1 + \alpha t)$, where α is the coefficient of linear expansion.

The new area will be

$$l_1 l_2 (1 + \alpha t)^2 = l_1 l_2 (1 + 2\alpha t + \alpha^2 t^2)$$

and if powers of α above unity are ignored, the new area is

$l_1 l_2 (1 + 2\alpha t) = $ original area $(1 \times 2\alpha \times$ change in temperature)

$= $ original area $ + (2\alpha \times$ original area $\times$ change in temperature)

The bracketed quantity is the change in area.

2α can be considered to be the coefficient of expansion of area of the material. This coefficient is of course an approximation. A similar approximation can be made for volume expansion, giving a coefficient of expansion of volume approximately equal to 3α.

Example 5.12

A steel cube of side 50 mm at 15 °C is heated to 85 °C. The coefficient of linear expansion of steel is 0.000 01/°C. Calculate the increased side length, surface area and volume.

Solution

Increased side length $= 50(1 + 0.000\,01 \times 70)$

where 70 is the change in temperature (°C)

$$= 50.035 \text{ mm}$$

Increased area $= $ (original area + increase in area) $\times 6$

(there are six faces to a cube) and since

increase in area = original area $\times$ change in temperature $\times$ coefficient of expansion of area

and the coefficient of expansion of area may be approximated to *twice* the coefficient of linear expansion, then

increase in area $= 2500 \times 70 \times 2 \times 0.000\,01 \text{ mm}^2$
$$= 3.5 \text{ mm}^2 \text{ per face}$$

increased area (total new area) $= 2503.5 \text{ mm}^2$ per face

therefore

total area $= 6 \times 2503.5 \text{ mm}^2$ of the cube as a whole
$$= 15\,021 \text{ mm}^2$$

It is interesting at this point to compare this approximate result with the accurate result obtained by squaring the new side length and multiplying by the number of sides. The new side length was determined to be 50.035 mm and thus the new side area is 50.035 $\times$ 50.035 which equals 2503.5012 mm^2, giving a total area, of 6 $\times$ this figure, as 15 021.007 mm^2, which is the correct area. The error is 0.000 04 per cent! (determined by electronic calculator).

Increase in volume = original volume $\times$ coefficient of expansion of volume $\times$ change in temperature

and the coefficient of expansion of volume may be taken as *three* times the coefficient of linear expansion, so that

$$\text{increase in volume} = 50^3 \times 3 \times 0.000\,01 \times 70$$
$$= 262.5 \text{ mm}^3$$
$$\text{increased volume} = \text{original volume} + \text{increase in volume}$$
$$= 125\,000 + 262.5$$
$$= 125\,262.5 \text{ mm}^3$$

The correct result obtained by cubing the new side length gives $50.035^3/\text{mm}^3$, that is, $125\,262.68 \text{ mm}^3$ and on this occasion the error in the approximation is 0.000 14 per cent. (It should be noted that such approximations, although derived using areas of squares and volumes of cubes, are applicable to bodies of any shape.)

EFFECT OF HEAT ON LIQUIDS AND GASES

Solids have a defined shape, and in considering the effect of heat on them we consider coefficients of linear expansion and of expansion of area and of volume. A liquid will change its shape and flow until it meets the side walls of the container in which it is situated. Consequently here we are not concerned with expansion of area or with linear expansion but only with volume change. The coefficient of cubical expansion (volume expansion) remains reasonably constant for a liquid over a defined temperature range and the range is usually quoted when making calculations. Water, for example, has a negative coefficient over the range 0 °C to 4 °C, so that it shrinks with increasing temperature and becomes more dense. This is why ice forms on the top of water, since the water at the slightly higher temperature than zero sinks and the colder water remains at the top. Gases behave differently to both liquids and solids, and expand to meet all container walls whatever the temperature, thus changing their volume and pressure as they do so. Pressure, volume and temperature of a gas are related by the general gas law.

ASSESSMENT EXERCISES

Long Answer

(Unless otherwise stated take the specific heat capacity of water as 4.2 kJ/kg °C)

5.1 Define specific heat capacity and relative specific heat capacity. Calculate the heat required to change the temperature of 12 kg of water from 20 °C to boiling point.

5.2 A block of metal of mass 33 kg and relative specific heat capacity 0.4 is heated so that its temperature changes from 350 K to 450 K in 3 min. Determine the average power output of the heater.

5.3 Define relative specific heat capacity of a material. To change the temperature of 5 kg of a certain material by 73.2 °C the total energy required is found to be 350 kJ. Calculate the relative specific heat capacity of the material.

5.4 Define thermal capacity and water equivalent of a body. Determine the thermal capacity and water equivalent of 15 kg of the material in question 5.3.

5.5 A body of mass 150 kg consists of copper and steel in unequal proportions. The thermal capacity of the body is 21 kJ/K. The relative specific heat capacities for copper and steel are 0.09 and 0.12 respectively. Calculate the ratio of copper to steel in the body.

5.6 A copper container with a water equivalent of 0.01 kg holds 0.045 kg of water at a temperature of 20 °C. A body of mass 0.02 kg at a temperature of 100 °C is immersed in the water. The relative specific heat capacity of the body material is 0.15. Find the final temperature of the water.

5.7 A metal bar at a temperature of 450 °C is immersed in a tank of oil at a temperature of 30 °C; the mass of the oil is 1.2 t; the final temperature of the oil and bar is 47 °C, the relative specific heat capacities of the metal and oil are 0.12 and 0.6 respectively. Find the mass of the metal bar.

5.8 In an experiment involving the immersion of a heated body in a calorimeter the following observations were made

mass of body 0.01 kg

mass of water in calorimeter 0.025 kg

temperature of water: before immersion 20 °C

after immersion 23.5 °C

temperature of body before immersion 70 °C

water equivalent of calorimeter 0.01 kg

heat lost to surroundings 120 J

Find the relative specific heat capacity of the body material and the water equivalent of the body.

5.9 A mass of 0.1 kg of ice at -6 °C is added to 0.8 kg of water at 15 °C contained in a calorimeter, water equivalent 0.025 kg. The relative specific heat capacity of ice is 0.45 and the latent heat of fusion is 335 kJ/kg. Calculate the final temperature of the water.

5.10 A calorimeter with a water equivalent of 0.01 kg contains 0.1 kg of water at 21 °C. Steam is passed into the water until the mass of the water increases to 0.102 kg. The latent heat of evaporation of water is 2257 kJ/kg. Calculate the final temperature of the water.

5.11 A 3-kW electric furnace is used to smelt metal from an initial temperature of 20 °C. The melting point of the metal is 200 °C and its latent heat of fusion is 40 MJ/kg. The specific heat capacity of the metal is 210 J/kg °C. Assuming no power loss or heat loss to surroundings, determine the mass of metal smelted per hour.

5.12 A bar of length 2.7 m at 17 °C is heated to 35 °C at which time the length has increased by 1.05 mm. Calculate the coefficient of linear expansion of the material of the bar.

5.13 Determine the expansion of a metal bar at 50 °C if the coefficient of linear expansion of the bar material is 0.000 017/°C and the bar length is 4 m at 20 °C.

5.14 A metal cube of side 75 mm at 20 °C is heated to 70 °C. The new side length is then 75.06 mm. Calculate the coefficient of linear expansion of the metal.

5.15 A length of heater pipe is 22 m long and contains water at 15 °C. The temperature of the water is increased to 75 °C. The coefficient of expansion of the pipe material is 0.000 016/°C. Find the final length of the pipe.

5.16 A square sheet of iron has an area of 0.07 m² at 10 °C. Determine the area at 50 °C. (Coefficient of expansion of iron is 0.000 012/°C.)

5.17 Calculate the heat energy required to raise the temperature of 0.8 kg of water from 20 °C to boiling point and then evaporate 25 per cent of the water. (Latent heat of evaporation is 2257 kJ/kg)

5.18 A calorimeter contains 0.1 kg of water at 20°C. When 0.005 kg of ice at 0 °C are added to the water its temperature falls to 18 °C. Calculate the water equivalent of the calorimeter. (Latent heat of fusion of ice is 335 kJ/kg)

Short Answer

5.19 Define 'thermal capacity' of a body.

5.20 What is the 'water equivalent' of a body?

5.21 Calculate the heat energy required to change the temperature of 2 kg of water by 20°C.

5.22 Determine the thermal capacity of a body of mass 5 kg and relative specific heat capacity 0.8.

5.23 Determine the water equivalent of 1 kg of material having a relative specific heat capacity of 0.3.

5.24 A bar of metal is immersed in water at 20°C. The water temperature rises to 35°C. Calculate the heat absorbed per kilogramme mass of the water.

5.25 The latent heat of fusion of ice is 335 kJ/kg. Calculate the heat required to melt 10 kg of ice without change in temperature.

5.26 The total heat absorbed by 10 kg of water evaporating without change of temperature is 22.57 MJ. What is the specific enthalpy of evaporation of water?

5.27 Calculate the increase in length per metre of a bar of metal having a coefficient of linear expansion of 0.000 018 / °C when the bar temperature is increased by 10°C.

5.28 Define 'coefficient of linear expansion'.

Multiple Choice

(Where required take the specific heat capacity of water = 4.2 kJ/kg °C)

5.29 The specific heat capacity of a material is
   A. the heat energy required to change the temperature of a body by one degree   B. the heat energy required to change the temperature of water by one degree   C. the change in temperature of unit mass of material per unit heat energy applied   D. the heat energy required to change the temperature of one unit of mass of a material by one degree

5.30 The heat energy required to change the temperature of a body is equal to
   A. specific heat capacity × mass × change in temperature   B. specific heat capacity × change in temperature   C. relative specific heat capacity × mass × change in temperature   D. relative specific heat capacity × specific heat capacity of water × change in temperature

5.31 The heat required to change the temperature of 5 kg of water by 15°C is equal to
   A. 21 kJ   B. 17.86 kJ   C. 75 kJ   D. 315 kJ

5.32 The heat energy required to change the temperature of a

20 kg mass by 17°C is equal to 850 kJ. Its relative specific heat capacity is
   A. 10.12   B. 10.5   C. 0.59   D. 2.5

5.33 The water equivalent of a body is equal to
   A. mass of body × specific heat capacity of body material   B. mass of body × relative specific heat capacity of body material   C. mass of body × specific heat capacity of water   D. specific heat capacity of body material × specific heat capacity of water

5.34 The thermal capacity of 7 kg of copper, if its relative specific heat capacity is 0.09, is
   A. 21 kJ/°C   B. 0.38 kJ/°C   C. 0.63 kJ/°C   D. 2.65 kJ/°C

5.35 The water equivalent of the copper mass in question 5.34 is
   A. 21 kg   B. 0.38 kg   C. 0.63 kg   D. 2.65 kg

5.36 A heated metal bar at 400 °C is immersed in oil at 20 °C. The oil temperature rises to 39 °C. The final temperature of the bar is
   A. 39 °C   B. 381 °C   C. 219.5 °C   D. 20 °C

5.37 Latent heat is
   A. the heat energy required when a body changes its temperature by one degree   B. the heat energy required when a body changes state   C. the heat energy given out when a liquid evaporates   D. the heat energy absorbed or released when a body changes state without changing temperature

5.38 The heat required when 0.2 kg of ice change state at 0 °C, assuming the specific enthalpy of fusion of ice to be 3.35 kJ/kg, is equal to
   A. 398.8 kJ   B. 281.4 kJ   C. 67 kJ   D. 1675 kJ

5.39 When a piece of metal expands due to heat the new length is equal to
   A. original length × coefficient of linear expansion × change in temperature   B. coefficient of linear expansion × change in temperature   C. original length × coefficient of linear expansion

   D. original length (1 + coefficient of linear expansion × change in temperature)

5.40 A metal bar of length 2.5 m is heated through a temperature change of 60 °C. The coefficient of linear expansion of the metal is 0.000 012 / °C. The length of the bar at the increased temperature is

   A. 2.5018 m   B. 0.0018 m   C. 2.500 72 m   D. 2.500 03 m

6 Waves

OBJECTIVES

All the objectives should be understood to be prefixed by the words
'The expected learning outcome is that the student . . .'

C9 Describes waves and their behaviour.
   9.1 Lists simple examples of wave motion.
   9.2 Explains, using a simple diagram, the meaning of
       (a) wavelength and (b) frequency.
   9.3 States the unit of frequency as the hertz.
   9.4 Solves simple problems using $v = f\lambda$
   9.5 Describes experiments which show that waves reflect and
       refract.
   9.6 Describes sound as a pressure wave.
   9.7 States that sound reflects and refracts.
   9.8 States that sound has a finite velocity, the value of which
       depends on the medium.
   9.9 States that sound is produced as a result of vibration and
       that the sound frequency depends on the form of the
       vibrator, for example, size of tuning fork, length of air
       column.

Most of us at one time or another have watched the sea and seen the tide coming in. Anyone who has, has seen for himself that the waves of water carry energy, particularly if the tide is assisted by the wind. The energy of the tide comes from the gravitational force of the Moon acting on a large mass of water and also from the wind, if this is present. On a small scale the energy of the sea is used to move buoys, which carry some kind of audible warning of hazards ahead for shipping; on a larger scale much thought is at present being devoted to the possibility of using tidal energy for electricity-generation.

Here, then, is an everyday example of energy being conveyed by wave motion. In fact all the energy that we receive from the Sun is conveyed by waves, although in this case the waves are not as immediately obvious as those of the sea. They are *electromagnetic waves*, whereas the sea waves are movements of matter—liquid matter in this case. Waves may be transmitted through any kind of medium—solid, liquid or gas—also, if they are electromagnetic, they may be transmitted through a vacuum. The same theory of wave motion applies to waves in and out of a medium, the medium itself having an effect on the waves but not altering the basic principles of wave motion.

A wave is defined in a dictionary as 'a motion to and fro or up and down', 'a periodic variation of a physical quantity' and 'a series of advancing impulses set up by a vibration'. As is often the case, a dictionary definition is a good place to start a closer

examination of what exactly is meant by a particular term in physical science. All three of these definitions are indicated in our first example of wave generation.

Imagine a rope tied at one end to a wall and the other end being held in someone's hand, as shown in figure 6.1. If the hand is now moved up and down a wave travels down the rope towards the fixed end. Each part of the rope remains in its own vertical plane but the wave travels along the rope. Each part in turn follows the initial hand movement and moves first in one direction and then in the other about a centre position. At any one time one part is at one particular place in its path of movement, while the part of the rope next to it is in a slightly different place in its path of movement. The over-all effect is that, for the situation shown, the movement of the rope particles is vertical and the direction of travel of the wave is horizontal. This kind of wave is called a *transverse* wave. In this example we see that an up-and-down motion causes a periodic variation of a physical quantity (displacement of each part of the rope from its centre position) and sets up a series of advancing impulses along the rope.

A second example, not as easily illustrated but shown as well as possible in figure 6.2, is that of a coiled spring lightly supported at

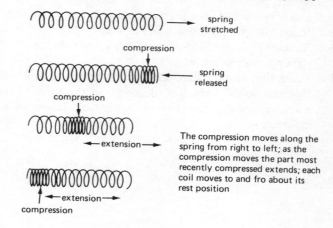

Longitudinal Wave Generation

Figure 6.2

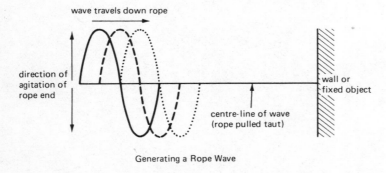

Generating a Rope Wave

Figure 6.1

each end, the spring axis being horizontal. If one end is pulled back, so that that part of the spring near the hand is extended, and then released, a spring action now causes compression and a wave moves along the spring, each part of the spring being first compressed then extended as the energy is transmitted along it. This wave is called a *longitudinal* wave, its characteristic being that the movement of the spring particles is in the *same* direction as the direction of travel of the wave. Again we have a periodic variation setting up a series of advancing impulses; this time the periodic variation is caused by a to-and-fro rather than an up-and-down movement.

Water or liquid waves such as those caused by a stone dropped in a pool are complex waves and contain both transverse and longitudinal waves.

Our examples so far illustrate waves within a medium—a liquid or solid. Sound waves, which will be considered again later, are another example of waves transmitted within a medium—in this case, liquid, solid or gas. To return now to electromagnetic waves, let us have a closer look at what causes these to be set up and how our dictionary definition applies on this occasion.

An electomagnetic wave carrying energy is set up whenever a changing electric field acts at right-angles to a changing magnetic field. Electric and magnetic fields are discussed in more detail in other chapters; for the moment it is sufficient to say that an electric field is set up by a voltage and a magnetic field by an electric current. An electric field exerts a force on electrically charged particles, whereas a magnetic field exerts a force on particles of magnetic material. When electromagnetic radiation occurs, a graph plotting field strength (electric or magnetic) against distance from the point of radiation has the same wave shape we saw earlier in figure 6.1; these field strengths are continually changing, so that the electromagnetic wave moves away from the source in much the same way that the changing rope positions caused the wave to move away from its source (in that case, the hand moving the rope up and down). So again we have a periodic variation (of current, voltage and field strength) setting up a series of advancing impulses.

Electromagnetic radiation may produce the sensation of heat or light; it may be used for radio or radar transmission or it may produce cosmic rays, X-rays, ultraviolet or infrared rays, depending on certain characteristics of the radiation. These characteristics include *frequency* and *wavelength*. These terms will now be defined.

CHARACTERISTICS OF A WAVE

All waves so far considered have a shape as shown in figure 6.3. Here a 'disturbance' is plotted against 'distance'. The disturbance may be a measure of displacement of matter, such as a rope, a spring or a liquid, from some rest position, or it may be a measure of field strength in the case of an electromagnetic wave. The 'distance' axis refers to the distance moved by the wave away from its source. The following definitions are very important.

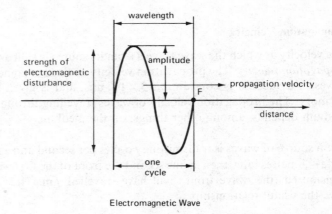

Electromagnetic Wave

Figure 6.3

Cycle

A complete vibration or alternation of the disturbance from rest to a maximum in one direction through rest to a maximum in the opposite direction is called one *cycle*.

Frequency

The number of cycles occurring per second is called the *frequency*. One cycle per second is called one hertz, abbreviated Hz. The symbol for frequency is f.

Periodic Time

The time taken per cycle is called the *periodic time*. The symbol is T. A little thought will show that periodic time = 1/frequency or, using symbols, $T = 1/f$.

Wavelength

The distance separating corresponding points on adjacent cycles is called the *wavelength* and is measured in metres. The symbol for wavelength is λ.

Propagation Velocity

The velocity at which the wave travels from its source is called the *propagation velocity*. The propagation velocity of electromagnetic waves in a vacuum is constant, at 3×10^8 m/s, and is given the symbol c. The propagation velocity of waves travelling through a medium depends, among other things, on the medium.

If a source of waves is transmitting f cycles per second and each cycle is λ metres long, then in one second the front of the first wave transmitted (the 'wave front') will have travelled f m. Thus we have the useful relationship

propagation velocity = frequency × wavelength

and we see that for constant propagation velocity, as in the case of electromagnetic waves in a vacuum, the higher the frequency the shorter the wavelength.

THE ELECTROMAGNETIC SPECTRUM (Table 6.1)

The complete range of electromagnetic waves varies from very low to very high frequencies (very long to very short wavelengths) and is called the electromagnetic spectrum. Details of the spectrum are given in table 6.1. Broadly speaking, the range may be divided into audio waves, radio waves (including television and radar waves), heat waves, light waves and the upper region of frequency which yields ultraviolet rays, X-rays and cosmic rays (the word wave is replaced by ray in this region).

Audio waves are those which, when received and processed by suitable equipment, give rise to sound waves that can be heard by the ear. Audio waves as such cannot be heard, because they are solely an electromagnetic disturbance; however, if alternating currents changing at frequencies within the audio range are fed to a loudspeaker, sound waves are generated, and these can be heard. Sound waves as distinct from electromagnetic audio waves will be considered in more detail later in the chapter.

Electromagnetic waves at audio frequencies are difficult to propagate and it is usual to use higher frequency waves in the radio frequency band to carry audio intelligence, using a process called modulation. The range of frequencies used as carriers, sometimes called *hertzian* waves after their discoverer, extends from just above the audio range at about 20 kHz to just below visible light, the infrared region beginning at around 750 GHz. Above the radio wave region there are radiant heat waves, light waves and the ray region. The visible light spectrum is quite small and it should be noted that, unlike electromagnetic audio waves, *electromagnetic waves at light frequencies* do produce the sensation of light directly on the sensory organ, the eye.

As the frequency is further increased above the light region the waves become more penetrating and can cause physical changes in the human body. Ultraviolet radiation causes skin changes which, in a mild form, result in tanning of the skin and, in a severe form, can result in burning. The ability of X-rays to penetrate human tissue is well known, and intense forms of radiation at these frequencies are used to destroy living matter. The effects of prolonged exposure to cosmic rays, from which earthbound

Table 6.1   The Electromagnetic Spectrum

	Frequency (Hz)	Wavelength (m)	
	3.0×10^{22}	10^{-14}	
	3.0×10^{20}	10^{-12}	Cosmic rays
Gamma rays	5×10^{19}	6×10^{-12}	
	1.5×10^{18}	2×10^{-10}	
	2.5×10^{16}	1.2×10^{-8}	X-rays
Ultraviolet rays	3×10^{15}	10^{-7}	
	7.5×10^{14}	4×10^{-7}	
	3.75×10^{14}	8×10^{-7}	Light waves
Infrared (heat) waves	3×10^{12}	10^{-4}	
	7.5×10^{11}	4×10^{-4}	
Radar TV	8.9×10^{8}	3.37×10^{-1}	These are sample
VHF radio bands	4.7×10^{8}	6.38×10^{-1}	frequencies within
	5.4×10^{7}	5.55	this band
Short wave	1.6×10^{6}	187.5	
Medium long waves	2×10^{4}	1.5×10^{4}	Limit of
	20 Hz	1.5×10^{7}	human ear

In the middle of the table (Hertzian (radio) waves) spans from Infrared to Short wave.

Notes 10^{12} means 1 000 000 000 000 (that is, 1 followed by 12 zeros)

10^{-12} means 0.000 000 000 001, that is, zero followed by 11 $(12-1)$ zeros; in other words there are $(12-1)$ zeros following the decimal point

humans are protected by the atmosphere, are still being investigated during lunar and other extra-terrestrial missions.

REFLECTION AND REFRACTION

As was stated earlier, the propagation velocity of any wave depends on the medium through which it is travelling. Certain waves—for example, sound—cannot travel through a vacuum and so the propagation velocity here is zero. Electromagnetic waves have a constant propagation velocity in a vacuum, but their velocity too is subject to change on entering a medium and in certain cases it may be zero, or almost zero. This fact is used to prevent light, and to prevent or reduce radio or other waves, passing into areas where it is not required (dark-rooms, interference screening, X-ray screening, etc.).

If a wave enters a medium *normally*, that is, at right-angles to its surface, its velocity may be reduced or increased but its direction remains unchanged. If the wave approaches the medium surface at any other angle, there is a change of direction, because the part of the wave front which meets the surface first changes its velocity before those parts of the wave front which arrive later. If the velocity in the medium is less than that outside the medium, the angle between the surface normal (that is, the line drawn perpendicular to the surface) and the direction of travel of the wave is *reduced*. If the velocity in the medium is greater than that outside the medium, this angle is increased. This is shown diagrammatically in figure 6.4a. The change in wave direction is called *refraction*.

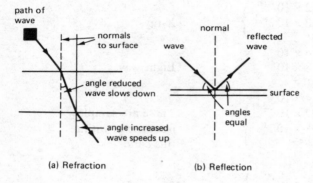

(a) Refraction (b) Reflection

Figure 6.4

When a wave meets a medium surface in which its velocity will change, some of the energy may be *reflected* and a reflected wave is set up, as shown in figure 6.4b. The approaching or incident wave and the reflected wave are symmetrical with the surface of the medium, as shown in the figure. If none of the wave energy enters the medium *total internal reflection* is said to occur, and this effect is used, among other instances, in guiding light along optic fibres into otherwise inaccessible places. The same effect in the case of sound is noticable in the Whispering Gallery at St Paul's Cathedral in London, where repeated total internal reflection of sound occurs, and this, together with the fact that the gallery dimensions are such that the sound is confined near to the surface, makes even a whisper audible.

SOUND WAVES

Sound is a *pressure* or *compressional* wave produced in a similar way to that of the spring which was mentioned earlier. Vibration of the source, that is, an actual movement to and fro or up and down, alternately *compresses* and *rarefies* the matter next to the source, and this compression and rarefaction is passed along from particle to particle along the path of the sound (see figure 6.5). Hence a vacuum or a near-vacuum has insufficient particles of matter to transmit the wave and so sound cannot travel in such conditions. This fact is well illustrated by placing an electric bell inside an inverted bell jar and evacuating the jar. As the air becomes more and more rarefied the sound dies but we can see that the bell is still ringing by the movement of its clapper.

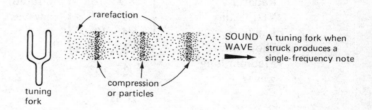

Figure 6.5

Sound frequencies detected by humans range from about 20 Hz to 20 kHz, but both limits vary considerably from person to person, being dependent on age, health, and so on. Certain animals, noticeably dogs and bats, hear higher frequencies (hence 'dog whistles' inaudible to humans); frequencies immediately above 20 kHz but below normal radio frequencies are classified as *ultrasonic*. Ultrasonic waves are very useful in intruder alarms, because they are virtually undetectible by humans; they are also used in industrial cleaning processes.

Sound waves have a velocity dependent on medium; in air it is approximately 333 m/s depending on temperature. Supersonic velocities are often expressed as a ratio called *Mach number*, where

$$\text{Mach number} = \frac{\text{velocity of body}}{\text{velocity of sound in the same medium}}$$

'Mach 1' means that a body is travelling at the velocity of sound, 'Mach 2' that it is travelling at twice the velocity of sound, and so on.

When a sound wave is produced by vibration, the frequency of the sound is determined by the vibrating source characteristics, for example, the shape and size of, say, a 'tuning fork' (used to provide a single-frequency sound source), the tension and mass/length of a musical instrument string, and the length and diameter of pipes used in organs and other wind instruments.

Note Assessment exercises for this chapter appear at the end of chapter 7.

7 Light

OBJECTIVES

*All the objectives should be understood to be prefixed by the words
'The expected learning outcome is that the student . . .'*

H21 States and applies the properties of light rays.

21.1 States the laws of reflection for plane mirrors.

21.2 Draws diagrams of ray paths for a plane mirror using the laws of reflection.

21.3 Draws, from observations, the paths of light rays incident on and passing through a parallel-sided glass block.

21.4 Defines refraction.

21.5 Names and sketches the shapes of bi-convex and bi-concave lenses.

21.6 Sketches the paths of rays passing through such lenses, from observation.

21.7 Sketches the path of a parallel beam of light passing through a lens parallel to the axis.

21.8 Defines focal length.

21.9 Explains the formation of virtual and real images in terms of rays.

LIGHT RAYS

The sensation which we recognise as light is caused by electromagnetic waves with a frequency between 375 and 750 terahertz (one terahertz is a million million cycles per second) and a wavelength between 760 and 380 nanometres (a nanometre is one-thousandth of a millionth of a metre). Radiations lying outside these limits have no visual effect on the human eye; those with wavelengths just longer give the sensation of heat and are called infrared, and those with wavelengths just shorter cause tanning of skin and, if intensive, can cause damage to body tissue; these latter rays are called ultraviolet. The light spectrum itself is divided into seven regions, each region having a different effect on the eye in terms of what we call 'colour'. These regions are red, orange, yellow, green, blue, indigo and violet. Red light is at the higher-frequency, longer-wavelength end of the spectrum and violet light is at the lower-frequency, shorter-wavelength end. White light is made up of all these radiations and, by a suitable arrangement of glass prisms and lenses, it can be split into its individual components. Electromagnetic waves in general are considered in chapter 6; in this chapter we shall take a closer look at the control of light using prisms and lenses and also mirrors.

In determining the effect of materials on light waves it is useful to use the idea of a 'ray'. An assumption is made that light travels in straight lines (a close approximation) and lines are drawn from the light source indicating the direction of travel of the wave. These lines are called rays and the outer rays of a beam contain the beam as a whole, as shown in figure 7.1. The figure shows a parallel beam, where the outer rays never meet or move further apart, a convergent beam, where the outer rays eventually meet at some distance from the source, and a divergent beam, where the outer rays move further apart as they travel from the source. Using the processes of *reflection* and *refraction* of light rays, *images* of any object may be produced. An image in this sense is a visual impression of the object. Images produced by light rays may be larger or smaller than the original object and the study of the production of such images is called *geometrical optics*. Geometrical optics is used in engineering in the design, development and production of optical systems for use in cinema and television

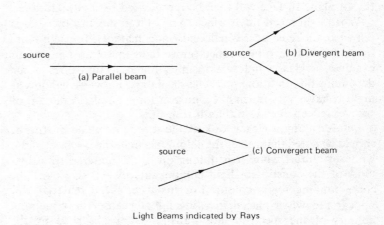

(a) Parallel beam
(b) Divergent beam
(c) Convergent beam

Light Beams indicated by Rays

Figure 7.1

equipment, indicating instruments and illumination schemes, to name but a few.

REFLECTION AND REFRACTION

When a light ray is *reflected* from a surface of a material it does not pass through the surface but returns from it either in the same or in a different direction to that in which it approached the surface. When a light ray is *refracted* by a material it passes through the material and its direction of travel is changed by the material. The change in direction is caused by the material affecting the speed of travel of the light waves.

Reflection at a Plane Surface

A plane surface is one which wholly contains every straight line joining any two points lying on it. Thus this page is a plane surface when it is lying flat but may not be so when it is turned, since a book page usually curves on turning. In diagrams any surface may be represented by the line which is seen when viewing one edge and thus a plane surface will then be represented by a straight line.

When light meets any surface some of it passes through and is refracted, the remainder is reflected. Some materials transmit more than they reflect, others reflect more than they transmit. As an example, a glass window transmits, but reflection does occur depending on conditions on each side of the window—a window at night behaves very much like a mirror, for example. A mirror, of course, reflects far more than it transmits.

Before taking a closer look at the theory of plane mirrors it should be noted that we see the colour of a material because, of the white or almost white light falling on it, the material transmits most of the constituent light components but reflects the light corresponding to its colour. For instance, a red material will appear red when placed in white light because the remaining colours of the spectrum are transmitted or absorbed by the material whereas the red light is not. Illuminating any material with light which is not purely of one colour may cause the material to appear to change its colour, since the degree of transmission or absorption and reflection depends not only on the colour of the components of light but at what points in the spectrum the colours making up the light are placed. The effects are also determined to a large degree by individual colour sensitivity of the eye and brain making the observation. A white material reflects all the colour components of white light but the reflection is irregular and does not occur in the same way as it does with a mirror surface. A mirror surface produces a regular or *specular* reflection as described below.

Figure 7.2 shows a parallel beam of light entirely reflected by a plane surface. Since the beam is parallel, all the rays approaching the surface, which are called *incident* rays, are at the same angle to the surface. Reflection of each ray is symmetrical, so that the angle made by the reflected ray with the surface is the same as that made by the incident ray with the surface. Usually the angles we consider are not those made with the surface but are those made with a line prependicular to the surface called the *normal*. As shown the angle of incidence and angle of reflection, both denoted by θ, are equal.

Figure 7.3 shows the formation of an image using a plane mirror. The object lies to the right of the mirror and two rays from any point on the object are drawn to the mirror surface. Each ray is

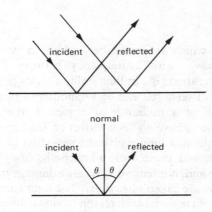

Reflection at a Plane Surface

Figure 7.2

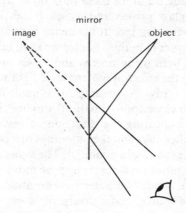

Image Formation in a Plane Mirror

Figure 7.3

reflected according to the laws of reflection described above. If now the reflected rays, which continue to diverge after reflection, are continued back behind the mirror, their continuations meet at a point as far behind the mirror surface as the object is in front. An eye observing the diverging rays, as shown, sees an image of the point lying behind the mirror and over all will see an image of the object as a whole.

Since each point has an image behind the mirror at the same distance as that point is in front of the mirror, those points on the object lying furthest away from the mirror surface will produce images lying furthest away behind the mirror. This produces the effect known as *lateral inversion*, as shown in figure 7.4. It is most obvious if a mirror image of writing is produced when the mirror writing appears 'backwards'.

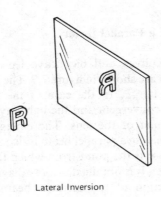

Lateral Inversion

Figure 7.4

A mirror image produced by a plane surface is the same size as the object and is upright (or erect), that is, the 'same way up' as the object, and, because it is situated at a point from which light rays appear to come, it is called a *virtual* (as opposed to a *real*) image. A real image is one which could be cast on to a screen as by film projector. Curved mirrors also produce images but they may be larger or smaller than the object, may be upright or inverted and may be real or virtual, depending on whether the mirror surface curves outwards towards the object (convex) or inwards away from the object (concave). A driving mirror is a convex mirror and gives a wide angle of view, although it is not as easy to judge

distances with such a mirror as with a plane mirror, since the images are not necessarily the same distance from the mirror as the objects they represent. Curved mirrors are also used as reflectors in headlamps, spotlamps and torches. Here the curve is a particular type of curve called a *parabola* and the reflector is called a parabolic reflector. Parabolic reflectors produce a parallel beam of light from a source producing a divergent beam.

Refraction

In a vacuum the speed at which light and all electromagnetic waves travel—the velocity of propagation—is constant at 3×10^8 metres per second. When light enters a medium, however, it slows down and the amount by which the velocity of propagation is reduced depends on the nature of the medium. In some substances it becomes virtually zero and these substances can be used to block out light waves. If a parallel beam of light is incident normal to a surface, that is, it enters at right-angles to the surface, then it slows down as it passes through the medium, but retains its original direction both in the medium and when it emerges from the other side. If the same beam is not normal to the surface then the part of the wave front which arrives first begins to slow down before the remaining parts of the wave front which arrive later, and the overall effect is to make the beam deviate from its original direction. If the beam is travelling from one medium into another, and the second medium slows the beam, then the direction of travel moves towards the normal. If the beam enters from one medium into a second, which allows a higher velocity of propagation, the direction of travel moves away from the normal. The change in direction is called *refraction*.

Refraction of a ray of light as it passes through a block with parallel sides is shown in figure 7.5. The incident ray is not normal to the entry surface and, assuming a medium which slows the light, the direction moves towards the normal as the ray enters. On emergence into the original medium the ray direction changes again, moving away from the normal this time and, because the exit surface is parallel to the entry surface, the direction after leaving is the same as that before entering. If the sides are not parallel, the change in direction still occurs but the directions of

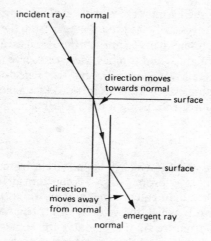

Refraction in a Parallel-sided Block

Figure 7.5

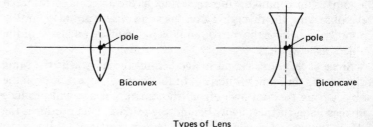

Types of Lens

Figure 7.6

travel before entering, while in, and after leaving the medium, are all different. Refraction of light is used to control direction, to focus beams and to produce images. To produce refraction we use a piece of optical apparatus called a *lens*.

LENSES

A lens is a piece of transparent material and usually has one or more curved surfaces. The most common shapes are the biconvex and biconcave, as shown in figure 7.6. The biconvex lens has two surfaces each curving outwards from the lens centre or *pole*. The biconcave lens has two curved surfaces curving inwards towards the pole. Other shapes, not illustrated, include the plano-convex, with one convex side and one plane side, the plano-concave, with one concave side and one plane side, the convex meniscus and the concave meniscus; the last two have two curved surfaces curved in the same direction, the difference being in the degree of curvature.

Effects of Lens on a Parallel Beam

The effects of biconvex and biconcave lenses on an incident parallel light beam are shown in figure 7.7. The pole of each lens is denoted by the letter P. In the case of the biconvex lens the emergent beam is convergent and the light rays meet at point F, called the *focal point* of the lens. The emergent beam from a biconcave lens receiving a parallel beam is divergent and the focal point, F, in this case is the point from which the divergent beam appears to originate. It is obtained in a ray diagram by continuing back in the direction of the incident beam source the lines representing the emerging divergent rays. In both cases the distance between the pole and the focal point, PF in the figures, is

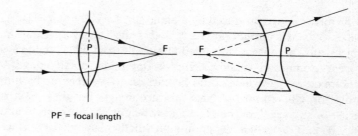

PF = focal length

Effect of Lens on Parallel Incident Rays

Figure 7.7

called the focal length. The focal length remains the same for these lens types regardless of the direction from which the incident beam approaches.

Image Production by Lenses

If an object from which light is emitted, either directly or by reflection, is placed at an appropriate distance from a lens, an image is produced. As stated earlier, images may be real or virtual, erect (upright) or inverted and may be larger, smaller or the same size as the object. The nature of the image depends specifically on the type of lens and on the relative position of the object with respect to the focal point and the lens pole, that is, on whether the object lies within the focal length, outside the focal length or at the focal point. Figures 7.8, 7.9 and 7.10 show the production of images by various lens types with the object placed at different distances from the lens. In all these constructions the technique is to take a specific point on the object and draw two rays, one from the point through the lens pole and the other parallel to the lens principal axis (this being as shown in the diagrams), to the lens and from there to emerge in the new direction as determined by the lens type. The direction of the ray from the point on the object through the lens pole remains unaltered as indicated.

Figure 7.8 shows image formation by a biconvex lens when the object is situated at a distance along the principal axis from the pole greater than the focal length. In all such cases the image produced is real and inverted. Its size, however, depends on the actual amount by which the distance between object and lens exceeds the focal length. If the distance is greater than *twice* the focal length, the image is larger; if the distance is smaller than twice the focal length (but still larger than the focal length) the image is smaller. If the object is placed at a distance along the principal axis exactly equal to twice the focal length the image is the same size as the object. Systems of this type producing smaller images are used in cameras, those producing larger images are used in enlargers or projectors and the type of system producing an image of the same size is used in an office copying-machine.

If the object is placed at the focal point the emergent rays are parallel and the image is said to appear at 'infinity', in other words,

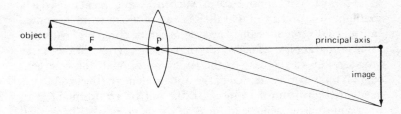

Image Formation by a Biconvex Lens: Object beyond Focal Point

Figure 7.8

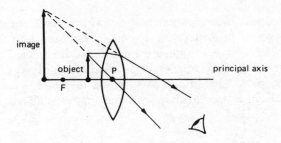

Image Formation by a Biconvex Lens: Object between Focal Point and Lens Pole

Figure 7.9

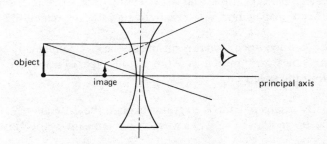

Image Formation by a Biconcave Lens

Figure 7.10

so far away that it cannot be seen. Such an arrangement is used in a spotlight to produce a parallel beam of light from a source emitting a divergent beam. The system can be compared with that using a parabolic reflector, in that both carry out the same function but, in the case of the lens arrangement, the emergent beam is on the opposite side of the optical element (the lens) to the incident beam from the object. In the mirror arrangement both emergent and incident beams are on the same side of the optical element, in this case, the mirror.

When the object is at a point on the principal axis between pole and focal point, that is, at a distance *less* than that of the focal length, the image produced is virtual, erect and magnified, as shown in figure 7.9. This arrangement is the principle of the magnifying glass.

A biconcave lens makes the rays which, when incident, are parallel to the principal axis, diverge on emerging and the image produced is always virtual, erect and reduced in size. The actual reduction depends on the exact position of the object relative to the lens pole. The situation is shown in figure 7.10 and the arrangement is used in spectacles and similar systems.

ASSESSMENT EXERCISES

Long Answer

7.1 Define the following characteristics of a wave: cycle, frequency, periodic time and wavelength. State the relationship between propagation velocity, frequency and wavelength.

7.2 Describe briefly what is meant by the electromagnetic spectrum, indicating typical uses of the various types of wave contained within the spectrum.

7.3 By means of simple diagrams explain the difference between reflection and refraction of a wave. What is meant by total internal reflection?

7.4 Using diagrams if necessary, distinguish between longitudinal and transverse waves. What type of wave is a sound wave and what is its main distinguishing characteristic?

7.5 By means of a diagram show how images are formed by a plane mirror. Explain the phenomenon of lateral inversion of such images. Why are such images called 'virtual'?

7.6 Describe, using diagrams, the following types of lens: biconvex, biconcave, plano-convex, plano-concave, convex meniscus and concave meniscus. Show how an image may be formed by any *two* of these lenses and indicate the exact nature of the image in each case.

7.7 Explain the meaning of the term 'focal length' when used in connection with lenses. By means of diagrams show how a biconvex lens forms an image of an object placed at a distance along the principal axis which is (a) greater than the focal length, (b) equal to the focal length, (c) less than the focal length. Give typical uses of such image forming optical systems in each case shown above.

7.8 With the aid of diagrams show how different types of lenses may be used (a) to produce a parallel light beam, (b) as a magnifying glass, (c) in spectacles. In each case state the type of lens used and the type of image produced.

Short Answer

7.9 What is the difference between a transverse and longitudinal wave?

7.10 State the relationship between propagation velocity, wavelength and frequency of a wave.

7.11 Calculate the wavelength of an electromagnetic wave of frequency 475 kHz if the propagation velocity is 3×10^8 m/s.

7.12 Calculate the frequency of an electromagnetic wave of wavelength 250 m if the propagation velocity is 3×10^8 m/s.

7.13 Give two uses of ultrasonic waves.

7.14 What is the difference between reflection and refraction of a light ray?

7.15 What is the distance between object and lens when a biconvex lens is used in copier?

7.16 If an object is placed at the focal point of a biconvex lens what kind of image is produced?

7.17 If an object is placed at a distance greater than the focal length of a biconvex lens along its principal axis what kind of image is produced?

7.18 State briefly how a virtual, erect and magnified image may be produced using a biconvex lens.

Multiple Choice

7.19 The periodic time of a wave of frequency 50 Hz is
A. 50 s   B. 20 ms   C. 0.2 s   D. 50 cycles/s

7.20 If the propagation velocity of an electromagnetic wave is 3×10^8 m/s, the wavelength of such a wave having a frequency of 1 MHz is
A. 300 m   B. 3×10^{14} m   C. 0.0033 m   D. 150 m

7.21 For a wave of constant propagation velocity
A. the higher the frequency the longer the wavelength   B. the lower the frequency the shorter the wavelength   C. the higher the frequency the shorter the wavelength   D. the frequency and wavelength are constant

7.22 Which of the following statements is true?
A. light waves have a higher frequency than X-rays   B. X-rays have a shorter wavelength than cosmic rays   C. ultraviolet rays have a higher frequency than gamma rays   D. gamma rays have a higher frequency than X-rays

7.23 If a wave centers a medium normally
A. its velocity always remains unchanged   B. its direction remains unchanged   C. its velocity always increases   D. its direction always changes

7.24 A velocity of Mach 0.5 is equal to
A. the velocity of sound   B. twice the velocity of sound   C. half the velocity of sound   D. one and a half times the velocity of sound

7.25 The image formed by a plane mirror is
A. virtual and inverted   B. real and inverted   C. virtual and erect   D. real and erect

7.26 For a biconvex lens, if the object is at a distance along the principal axis greater than twice the focal length, the image is
A. real, inverted and diminished   B. real, erect and larger   C. virtual, inverted and diminished   D. real, erect and diminished

7.27 For a biconvex lens, if the object is at a distance along the principal axis equal to twice the focal length, the image is
A. real, erect, same size   B. real, inverted, same size   C. virtual, erect, larger   D. real, inverted, smaller

7.28 For a biconvex lens, if the object is at a distance along the principal axis less than twice the focal length but greater than the focal length, the image is
A. virtual, inverted, diminished   B. real, erect, same size   C. real, inverted, magnified   D. virtual, inverted, magnified

8 Stress and Strain

OBJECTIVES

*All the objectives should be understood to be prefixed by the words
'The expected learning outcome is that the student . . .'*

A2 Describes the effect of forces on materials with different
properties.
2.1 Selects, from given examples of forces diagrams, those
components in (a) tension (b) compression (c) shear.
2.2 Defines elasticity.
2.3 States Hooke's law.
2.4 Solves simple problems involving Hooke's law.
2.5 Describes the difference between the behaviour of
(a) brittle and (b) malleable materials under the action of
a force.
2.6 Draws and labels a load–extension graph, for (a) ductile
and (b) brittle materials, using experimental data.

When forces are applied to a piece of material tending to stretch it, as shown in figure 8.1a, the material is said to be in *tension* and the forces are called *tensile forces*. If the forces are such that the material is tending to be squashed, it is said to be in *compression* and the forces are *compressive forces* (figure 8.1b). The length of a bar subjected to tensile forces will increase and when subjected to compressive forces it will be reduced. Sometimes the forces acting on a body are such that the body tends to separate into layers, one layer sliding over another. These forces are called *shear forces*. Figure 8.1c shows two plates bolted together and forces applied in opposite directions to the plates. In this case the holding bolt is being subjected to shear forces. All forces, tensile, compressive or shear, tend to cause distortion of the body on which they act. The resulting distortion depends not only on the magnitude of the force but also on the area on which or over which the forces act.

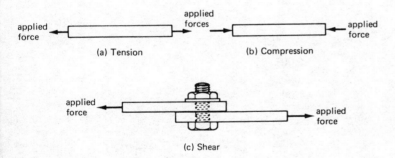

(a) Tension

(b) Compression

(c) Shear

Figure 8.1

STRESS

Whenever forces act on a body such that tension, compression or shear occurs, the body is in a state of *stress*. Stress is defined as the ratio of the applied force to the area on which the force acts. For tensile and compressive forces this area is the area at right-angles to the direction of action of the force. For shear forces the area is that of the layers over which the forces act. This is clarified below.

$$\text{Stress} = \frac{\text{applied force}}{\text{cross-sectional area of material}}$$

Stress is sometimes referred to as 'intensity of stress' and in these applications the applied force is called the *load*.

Example 8.1

A bar of circular cross-section, diameter 0.15 m, is subjected to a tensile force of 250 N. Calculate the intensity of tensile stress in the bar for this value of load.
Solution

$$
\begin{aligned}
\text{Cross-sectional area} &= \pi \times \text{radius}^2 \\
&= \pi \times 0.075^2 \\
&= 0.0177 \text{ m}^2
\end{aligned}
$$

$$\text{Load} = 250 \text{ N}$$

$$
\begin{aligned}
\text{Intensity of stress} &= \frac{250}{0.0177} \\
&= 14\,124.3 \text{ N/m}^2 \\
&= 14.12 \text{ kN/m}^2
\end{aligned}
$$

Note that the unit of stress is the same as that of pressure. The special name given to the newton per square metre is the *pascal*, symbol P.

Example 8.2

A mass of 150 kg rests on a support with a rectangular cross-section of dimensions 125 mm × 350 mm. Calculate the compressive stress in the support.
Solution

$$
\begin{aligned}
\text{Cross-sectional area of support} &= 125 \times 350 \\
&= 43\,750 \text{ mm}^2 \\
&= 0.043\,750 \text{ m}^2
\end{aligned}
$$

(dividing by 1 000 000)

Load = weight of body
= 150 × acceleration due to gravity
= 150 × 9.81
= 1471 N

$$\text{Compressive stress} = \frac{1471}{0.043\,750}$$

$$= 33\,622.86 \text{ N/m}^2$$
$$= 33.62 \text{ kN/m}^2 \text{ or kP}$$

Example 8.3

A bar of square cross-section, side length 25 mm, is subjected to a shear force of 75 N. Calculate the shear stress in the bar.
Solution

Bar area of cross-section = 0.025^2 m^2
Load = 75 N

$$\text{Shear stress} = \frac{75}{0.025^2} \text{ N/m}^2$$
$$= 120 \text{ kN/m}^2$$

Example 8.4

A circular hole of diameter 25 mm is to be punched out of a metal plate of thickness 5 mm. The force applied is found to be 87 kN. Calculate the shear stress required to cause fracture.
Solution   The area to be sheared is equal to the circle circumference multiplied by the thickness of the plate

area = 2π × radius × thickness
= 2π × 12.5 × 5
= 392.86 mm^2

$$\text{Shear stress} = \frac{\text{load}}{\text{area}}$$
$$= \frac{87}{392.86}$$
$$= 0.221 \text{ kN/mm}^2$$

STRAIN

The distortion of a body subjected to tensile, compressive or shear forces is measured in terms of a quantity called *strain*.

For tensile or compressive forces the body length changes and the strain is defined as the ratio change in length : original length. Both lengths are measured in the same units, so that strain is a pure number and has no units.

Example 8.5

Prior to any force being applied, the normal length of a steel bar is 0.45 m. When a tensile force is applied the length changes to 451 mm. Calculate the strain.
Solution

Change in bar length = 0.451 − 0.45
= 0.001 m

Bar length = 0.45 m

$$\text{Strain} = \frac{0.001}{0.45}$$
$$= 0.0022$$

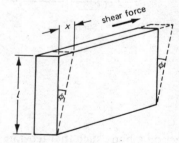

Figure 8.2

If a block such as that shown in figure 8.2 is subjected to a shear force so that it is distorted to the shape shown by the dashed lines, the shear strain is defined as length x divided by length l, that is, the linear displacement of a side in the direction of the line of action of the shear force, divided by the length of the side measured in a direction perpendicular to that of the line of action of the shear force. If the angle shown as ϕ in the figure is expressed in radians, then (from trigonometry)

$$\frac{x}{l} = \phi$$

so that the shear strain may also be defined as the angle, in radians, through which the side length moves from its normal position.

RELATIONSHIP BETWEEN STRESS AND STRAIN

Increasing the load (tensile, compressive or shear) and therefore the strain on a piece of material increases the distortion of the material and thus the stress. Up to a certain value of load the distortion is not permanent—when the load is removed the material returns to its original size or shape. The point at which this no longer applies is called the *elastic limit* of the material. If loads are applied beyond the elastic limit, then when they are removed some distortion will remain.

Provided that the distortion is kept within the elastic limit it can be shown that stress is directly proportional to strain. A graph plotting stress against strain would therefore be a straight-line graph, as shown in figure 8.3. (Beyond the elastic limit a different situation exists and this will be discussed further later.) The statement

the strain in a material is directly proportional to the stress which produces it, provided the strain is kept within the elastic limit

is known as *Hooke's law*.

The ratio of stress to strain is given a special name: for tensile or compressive stress, stress/strain is called the 'modulus of elasticity' or 'Young's modulus', symbol E; for shear stress the ratio stress/strain is called the 'modulus of rigidity', symbol G.

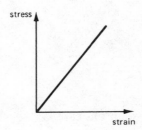

Figure 8.3

Since strain has no units the ratio stress/strain has the units of stress, namely those of force per unit area.

Example 8.6

A tensile force of 300 N is applied to a wire of circular cross-section, diameter 5 mm, and of normal length 2.5 m. The modulus of elasticity of the wire material is 197 kN/mm². Assuming the elastic limit is not exceeded, calculate the extension produced in the wire.
Solution

$$\text{Area of cross-section} = \pi \times \text{radius}^2$$
$$= \pi \times 2.5^2$$
$$= 19.64 \text{ mm}^2$$

$$\text{Load} = 300 \text{ N}$$

$$\text{Stress} = \frac{\text{load}}{\text{area of cross-section}}$$
$$= \frac{300}{19.64}$$
$$= 15.27 \text{ N/mm}^2$$

Now stress/strain is the modulus of elasticity, E, so that

$$E = \frac{\text{stress}}{\text{strain}}$$

and

$$\text{strain} = \frac{\text{stress}}{E}$$

$$= \frac{15.27}{197\,000} \quad \text{in this case}$$

Note that the units of stress have been kept as N/mm^2 because Young's modulus, E, is given in kN/mm^2. (In the denominator it is necessary to write 197 000 and not 197 so that E may be in N/mm^2, the same units as the stress.)

Hence strain = 0.000 08, and since strain = extension/normal length

$$\begin{aligned}\text{extension} &= \text{strain} \times \text{normal length} \\ &= 0.000\,08 \times 2500 \\ &= 0.194\ \text{mm}\end{aligned}$$

Example 8.7

A metal rod of square cross-section, side length 34 mm, is subjected to a compressive force of 125 kN. The length of the rod before compression is 3 m and after is 2.987 m. Calculate the value of Young's modulus for the material assuming the elastic limit is not exceeded.

Solution

$$\begin{aligned}\text{Area of cross-section} &= 34 \times 34 \\ &= 1156\ \text{mm}^2\end{aligned}$$

$$\begin{aligned}\text{Stress} &= \frac{125}{1156} \\ &= 0.108\ \text{kN/mm}^2\end{aligned}$$

$$\begin{aligned}\text{Change in length} &= 3000 - 2987 \\ &= 13\ \text{mm}\end{aligned}$$

$$\text{Strain} = \frac{13}{3000}$$

$$= 0.0043$$

$$\begin{aligned}\text{Young's modulus} &= \frac{0.108}{0.0043} \\ &= 25.12\ \text{kN/mm}^2\end{aligned}$$

Example 8.8

A bar of rectangular cross-section, width 15 mm, carries a load of 23 kN. The stress in the bar must not exceed 297 N/mm^2. Young's modulus for the bar material is 156 kN/mm^2. Determine (a) the bar thickness for maximum stress, (b) the strain on the bar at this loading.

Solution (a) Let x be the thickness of the bar so that

$$\text{area of cross-section} = 15x\ \text{mm}^2$$

$$\begin{aligned}\text{stress} &= \frac{\text{load}}{\text{area}} \\ &= \frac{23}{15x}\ \text{kN/mm}^2\end{aligned}$$

Maximum stress is 297 N/mm^2, that is, 0.297 kN/mm^2, thus

$$\frac{23}{15x} = 0.297$$

and

$$\begin{aligned}15x &= \frac{23}{0.297} \\ &= 77.44\end{aligned}$$

so that

$$x = 5.16\ \text{mm}$$

The bar thickness is 5.16 mm.

(b) Strain is equal to stress divided by Young's modulus, hence

$$\text{strain} = \frac{297}{156000}$$

(same units!)

$$= 0.0019$$

ULTIMATE STRESS AND FACTOR OF SAFETY

Figure 8.3 shows a typical stress–strain graph. Since stress is proportional to load, and strain is proportional to the extension of a piece of material subjected to a tensile force, then this graph has the same shape as a load–extension graph up to the point where the elastic limit is reached. If the load is increased beyond the elastic limit, the graph takes the shape shown in figure 8.4. Just beyond the elastic limit the extension per unit load suddenly increases (the yield point) and beyond this the material becomes *ductile*. At the point shown as maximum load, the extension in the material has reduced the cross-sectional area (since the volume remains the same and volume is equal to length × cross-sectional area) so that the effective stress is increased on two accounts: the load increase and the area decrease. If the maximum load is maintained, fracture of the material occurs. Reduction of the applied load at this maximum-load point, resulting in a reduction of stress or main-

taining of relatively constant stress, enables the material to be further extended but fracture occurs eventually. Figure 8.5 (courtesy of G. Cussons Ltd) shows a tensile testing machine used for

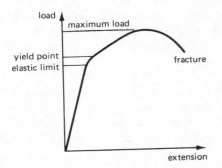

Figure 8.4

Figure 8.5

testing materials and obtaining a stress–strain or load–extension characteristics similar to that shown in figure 9.4.

The ultimate stress (also known as ultimate tensile strength or ultimate tensile stress) is the maximum stress which can be experienced by a material without fracture. It is defined as the maximum load divided by the *original* cross-sectional area of the material. It is not usually advisable to allow the stress to reach a value approaching the ultimate value; indeed, it is not normally desirable to go beyond the elastic limit, since beyond here any deformation is permanent. Accordingly, the allowable stress is limited by a *factor of safety*, which is the ratio between the ultimate stress and the allowable stress.

$$\text{Ultimate stress} = \frac{\text{maximum load}}{\text{original area}}$$

$$\text{Factor of safety} = \frac{\text{ultimate stress}}{\text{allowable stress}}$$

ASSESSMENT EXERCISES

Long Answer

8.1 Determine the tensile stress in a bar of circular cross-section, diameter 175 mm, subject to a tensile force of 375 N.

8.2 A bar of square cross-section, side length 150 mm, supports a body and the compressive stress is 150 kN/m^2. Find the mass of the body.

8.3 A bar of circular cross-section, diameter 52 mm, is subjected to a shear force of 120 N. Find the shear stress in the bar.

8.4 A circular hole of diameter 50 mm is to be punched out of a metal plate of thickness 7 mm. The shear stress required to cause fracture is 0.17 kN/mm^2. Determine the value of the force required.

8.5 A tensile force of 600 N applied to a wire of circular cross-section, 6 mm diameter, and of normal length 3 m produces an extension of 0.18 mm. The elastic limit is not exceeded. Find the modulus of elasticity of the wire material.

8.6 A metal bar of square cross-section, side length 50 mm, is subjected to a compressive force of 150 kN, the elastic limit not being exceeded. Young's modulus for the material is 30 kN/mm^2. Calculate the percentage change in the bar length.

8.7 A bar of circular cross-section, diameter 20 mm, carries a load of 30 kN. Young's modulus for the bar material is 170 kN/mm^2. Find the strain on the bar.

8.8 A 30-mm diameter punch and die are used to produce circular blanks from a sheet of thickness 3 mm and shear strength 400 N/mm^2. Calculate the force required and the stress in the punch.

8.9 A vertical load of 70 kN is applied to a cylindrical bar of diameter 70 mm. Determine the strain if Young's modulus for the material is 250 kN/mm^2.

8.10 A compressive load of 100 kN is applied to a metal tube of outside diameter 80 mm. The allowable stress is 85 N/mm^2. Determine the tube internal diameter.

8.11 A rectangular block of material has dimensions base length 400 mm, base width 25 mm, height 300 mm. The base is fixed to a surface and the opposite face is subjected to a load of 500 kN exerted parallel to the surface. The upper face moves 1.5 mm. Calculate the modulus of rigidity of the material.

8.12 The stress in a bar of material of circular cross-section is 20 N/mm^2 when a force of 750 N is exerted along the axis. Calculate the bar diameter. If the percentage extension of the bar is 0.005 when subjected to this force, calculate Young's modulus for the material.

8.13 A metal rod is subjected to a tensile load of 18 kN. If the length of the rod is not to change by more than 0.006 per cent, find the minimum cross-sectional area of the rod. (Take $E = 200 \, kN/mm^2$)

Short Answer

8.14 Define tensile stress.

8.15 Define tensile strain.

8.16 Calculate the stress when a compressive load of 500 N is applied to a bar of cross-sectional area 150 mm².

8.17 Calculate the strain if a bar of length 1.2 m is extended to a new length of 1.204 m.

8.18 Define Young's modulus.

8.19 Calculate the strain produced by a tensile stress of $15 \, kN/mm^2$ on a bar of material having $E = 195 \, kN/mm^2$.

8.20 The strain produced by a tensile load is 0.0015. The change in length is 1.5 mm. Find the original length of the bar.

8.21 State Hooke's law.

8.22 Define 'modulus of rigidity'.

8.23 Find the modulus of rigidity of a material if a shear stress of 1.2 kN/mm² produces a strain of 0.004.

8.24 State the SI units of (a) stress, (b) strain, (c) Young's modulus.

8.25 The modulus of elasticity of a certain material is 200 kN/mm². What strain would a tensile stress of 30 kN/mm² produce (provided the elastic limit were not exceeded)?

Multiple Choice

8.26 Mechanical stress is defined as
   A. applied force × cross-sectional area of material   B. length after force is applied ÷ original length   C. applied force ÷ cross-sectional area of material   D. length after force is applied × original length

8.27 A load of 500 N is applied to an area of 2 m². The stress
   A. is 250 N/m²   B. is 1000 N/m²   C. is 4×10^{-3} N/m²   D. cannot be determined without further information

8.28 When the length of a body changes due to the application of a tensile or compressive force the strain is defined as
   A. length after force is applied ÷ original length   B. change in length ÷ original   length   C. original   length ÷ change   in length   D. original length ÷ length after force is applied

8.29 Young's modulus is the ratio
   A. stress/strain   for   tensile   or   compressive   stress
   B. strain/stress for shear stress   C. stress/strain for shear stress
   D. strain/stress for tensile or compressive stress

8.30 The elastic limit is the point on the stress–strain curve
   A. at which a material subjected to stress breaks   B. below which strain is not proportional to stress   C. at which a material becomes ductile   D. beyond which some distortion remains on removal of the load

8.31 A tensile force of 400 N is applied to a wire of cross-sectional area 0.005 m² and produces a strain of 0.41×10^{-6}. Assuming the elastic limit is not exceeded the modulus of elasticity for the wire is
   A. 5.1×10^{-12} N/m²   B. 195.12×10^9 N/mm²
   C. 0.0328 N/m²   D. 195.12 kN/mm²

8.32 A bar of cross-sectional area 100 mm² is subjected to a shear force of 150 N. The shear stress is
   A. 15 kN mm²   B. 1.5 N/mm²   C. 0.67 N/mm²
   D. 15 kN/mm²

8.33 A disc of metal has a cross-sectional area of 150 mm^2 and the area of its circular edge (the circumference multiplied by the disc thickness) is 412 mm^2. The force required to punch out the disc from a larger piece of metal is 250 kN. The shear stress which caused fracture is

   A. 0.95 kN/mm^2 B. 0.44 kN/mm^2 C. 1.67 kN/mm^2
D. 0.61 kN/mm^2

8.34 The modulus of rigidity

   A. is the ratio stress/strain for a tensile stress. B. is also known as Young's modulus C. is the ratio stress/strain for a compressive stress D. is the ratio stress/strain for shear stress

8.35 When a tensile load is applied to a wire producing an extension the extension is equal to

   A. strain × original length B. strain ÷ original length C. stress ÷ Young's modulus D. stress × modulus of elasticity

9 Electricity

OBJECTIVES

All the objectives should be understood to be prefixed by the words 'The expected learning outcome is that the student . . .'

D10 Solves problems associated with simple electric circuits.

10.1 Selects and uses preferred unit prefixes in accordance with SI.

10.2 Uses the preferred symbols for electrical components when drawing circuit diagrams.

10.3 States that the unit of current is the ampere.

10.4 States that for continuous current a complete circuit is necessary.

10.5 Explains how current flows due to the existence of a potential difference (voltage) between two points in an electrical conductor.

10.6 States that the unit of potential difference is the volt.

10.7 Measures current using an ammeter and potential difference using a voltmeter.

10.8 Draws a graph of the relationship between potential difference and current using experimental data, for (a) a single resistor, (b) a non-linear component such as a lamp.

10.9 Defines resistance as the ratio of potential difference (voltage) across a resistor to the current through it and describes resistance as that property of a conductor that limits current.

10.10 States Ohm's law in terms of the proportionality of current to potential difference.

10.11 Solves simple problems using Ohm's law.

10.12 Describes the difference between series and parallel connections of resistors.

10.13 States that current is the same in all parts of a series circuit.

10.14 States that the sum of the voltages in a series circuit is equal to the total applied voltage.

10.15 Shows that for resistors connected in series the equivalent resistance is given by $R = R_1 + R_2 + R_3$.

10.16 Solves simple problems involving up to three resistors connected in series, including the use of Ohm's law.

10.17 States that the sum of the current in resistors connected in parallel is equal to the current flowing into the parallel network.

10.18 States that the potential difference (voltage) is the same across resistors in parallel.

10.19 Shows that for resistors connected in parallel the equivalent resistance is given by

$$\frac{1}{R} = \frac{1}{R_1} + \frac{1}{R_2} + \frac{1}{R_3}$$

10.20 Solves simple problems involving up to three resistors connected in parallel by the use of Ohm's law.

10.21 Sets up simple parallel and series circuits given a circuit diagram and measures current and voltage.

10.22 States the relationship between the resistance of a conductor and its length, cross-sectional area, and material.

10.23 States that resistance varies with temperature.

10.24 Compares the merits of wiring lamps in (a) series, (b) parallel.

D11 Identifies the three main effects of an electric current.

11.1 States examples of electric current being used for its magnetic effect.

11.2 States examples of electric current being used for its chemical effect.

11.3 States examples of electric current being used for its heating effect.

11.4 Identifies the effect being made use of in given specific cases, for example, electromagnet, electroplating, electric fire, fuse.

D12 Applies the power concept to electrical circuits.

12.1 States that the power produced in a circuit is equal to the product of potential difference and current.

12.2 Uses Ohm's law to show that $P = I^2R$.

12.3 Calculates the power dissipated in simple circuits.

I22 Solves problems involving resistance variation with temperature and resistivity.

22.1 Defines the temperature coefficient of resistance.

22.2 Solves simple problems involving the temperature coefficient of resistance.

22.3 Defines resistivity.

22.4 Solves simple problems involving resistivity.

22.5 Lists examples of conducting and insulating materials in common use.

It is difficult to define electric charge; probably the easiest thing to say is that we recognise it by its effects, which can be seen, heard and felt; occasionally, electric charge produces an effect which we can smell (the gas ozone).

The effects of electric charge have been known for thousands of years although it is only in comparatively recent times that research and study have produced the many hundreds of electrical devices and machines that we now have. The early Greeks were aware that when certain materials were rubbed they exhibited the property of exerting an attractive and repulsive force on certain other materials. Amber was one of these materials, probably the best known, and it is the Greek word for amber—*elektron*—which forms the stem of our words such as electricity, electron, electronics, electromagnetism, and so on.

There appear to be two kinds of electric charge. Ebonite rubbed with fur acquires one kind—now called *negative*—and glass rubbed with silk acquires another kind—now called *positive*. Bodies charged with opposite kinds of charge attract one another, bodies charged with the same kind of charge repel one another.

What is meant by acquiring charge? The structure of matter is very complex but modern atomic theory tells us that a fairly simple picture of matter is sufficient to explain a great number of things observed in science and in nature. The simple picture is that all matter is made up of minute particles, called protons—which exhibit the effects of the charge we call positive—electrons—which exhibit the effects of the charge we call negative—and neutrons—which have no electric charge. Protons repel each other and attract electrons, while electrons repel each other and attract protons. A body which acquires charge is either gaining or losing these particles. Most bodies in their normal state have equal numbers of negative electrons and positive protons and are electrically *neutral*. Gaining or losing the minute particles (usually electrons) means that the body will acquire either a negative charge or positive charge, respectively. It is now believed that rubbing a material with another—ebonite with fur, glass with silk—transfers electrons from one material to the other.

It should be appreciated, even at an early stage in studying electricity, that the theory which uses the idea of minute particles is a simplification. It has become apparent from studies in physics and electronics that sometimes these particles behave in a most unparticle-like manner! However, for the moment, the basic atomic ideas of the early twentieth century are quite sufficient to give us a mental picture of what is happening in our studies of electricity.

The Unit of Electric Charge

The amount of electric charge carried by an electron is very small, and the practical unit used in the SI System of units is equivalent to the charge carried by many millions of electrons, 6.3×10^{18} of them, to give an approximate figure. The unit of electric charge is called the *coulomb*, the symbol for coulomb being C.

ELECTROSTATICS AND ELECTRIC CURRENT

If a material is rubbed with another and acquires charge the charge so acquired may remain indefinitely. If the charge does not move it is said to be static and so the study of static charge is called electrostatics. If charge does move (carried by electrons or other particles) then we say that an *electric current* flows.

An electric current is a flow of electrically charged particles (usually, but not always, electrons). The unit of electric current is the *ampere*, symbol A, and one ampere is the electric current which transfers one coulomb of electric charge every second.

One ampere is equal to one coulomb per second.

Electromotive Force

For electric charge to be moved from one place to another the particles carrying the charge must be given energy. This energy may be given in any one of a number of different ways including the use of light, heat, chemical energy, magnetism (which we shall study shortly), and so on. The ability to give energy to charged particles is called *electromotive force*, abbreviated to e.m.f. This special name is widely used but, in a way, it is unfortunate, since the word force gives the wrong idea of the nature of the quantity;

electromotive force is not in fact a force at all. Since it is the ability to give energy to charged particles it is measured in energy units per unit of charge, that is, joules per coulomb. The unit of force (the newton) does not enter into the unit of e.m.f. at all. One joule per coulomb is given a special name, the *volt*, abbreviated V.

One volt is one joule per coulomb.

Electrical Resistance

As was stated earlier, all materials contain the basic atomic particles: neutrons, protons and electrons. The number of particles and their arrangement within the material give the material its characteristic properties and make it different from all other materials. To continue with the simple picture of atomic structure discussed earlier, all materials are assumed to be made up of *atoms*, each atom within a particular material being the same in structure and containing some of the three basic particles. The simple picture suggests that the atom consists of a nucleus containing neutrons and protons and a number of electrons which orbit continually in definite paths around the nucleus (see figure 9.1). The atom is held together by a complex pattern of forces some of which are due to the electrical nature of the particles.

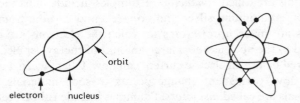

Basic Atomic Structure

Figure 9.1

All materials, then, consist of atoms, the atoms being interlinked to form the total material; different materials are different in their character because of the construction and content of their constituent atoms.

The electrons which orbit the atomic nucleus may be pulled from their orbit and even released from the material if sufficient energy is given to the material. The amount of energy required per electron depends on the material, and some materials require little energy to free electrons from the atoms, while others require a great deal of energy. Electrical resistance is a measure of the difficulty in freeing electrons from their orbits and establishing an electric current within a material.

The unit of electrical resistance is the *ohm*, symbol Ω (omega). A piece of material has an electrical resistance of one ohm if one volt is required to set up an electric current of one ampere. Another way of describing the ohm is the volt per ampere.

One ohm is one volt per ampere.

CONDUCTORS AND INSULATORS

It is possible to establish an electric current in any material provided that sufficient e.m.f., or *voltage*, is available. Certain materials require very little energy for the release of the electrons from the material atoms and the passage of electric current through them causes no significant change in their characteristics. Such materials are called *conductors*. Examples include most metals—notably aluminium, silver and copper—and carbon. Some materials are made up of atoms in which the electrons are closely bound to the nucleus and a large amount of energy per electron is required before an electric current can flow. Often, when a current does flow, permanent change occurs in such materials. These materials are called *insulators*. Examples include rubber, ceramics and plastics. If two similar-sized pieces of material are examined, one an insulator and one a conductor, it will be found that the conductor has a much lower electrical resistance than the insulator.

A third group of materials, called semiconductors, exists. The resistive characteristics of these materials lie between those of conductors and insulators. Semiconductors are widely used in electronics.

VOLTAGE—CURRENT GRAPHS

For most materials, the resistance of a piece of the material of fixed size at a fixed temperature remains constant. If a source of e.m.f. is attached to such a piece so that an electric current can flow, and if the values of e.m.f. and current are noted, it will be found that the voltage applied is directly proportional to the current flowing. A voltage–current graph will be linear, as shown in figure 9.2a. Certain special materials exhibit a resistance which depends on voltage applied and the graph may then be of the form shown in figure 9.2b. For this material the current suddenly increases at some value of voltage and the voltage–current graph is non-linear.

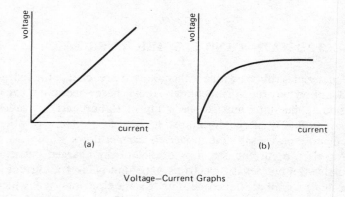

Voltage–Current Graphs

Figure 9.2

The statement that for most materials the current flowing is directly proportional to the applied voltage is often called Ohm's law, after the scientist who did a great deal of work investigating electrical characteristics of various materials. The essential fact to appreciate is that the ratio of voltage to current always gives a measure of the opposition of the material to electric current flow, and this opposition may be constant or may vary depending on voltage or some other variable.

RESISTORS

Components specially made to have a particular value of electrical resistance are called *resistors*. Many different types are available including carbon resistors, wire-wound resistors, metal oxide resistors, and so on. The value of the resistance of a resistor is either written on the component in figures or a special code using coloured rings is used (see figure 9.3). The resistor symbols shown in the figure are used in electrical circuit diagrams discussed below.

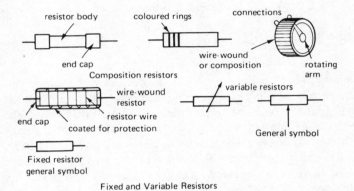

Figure 9.3

THE CONDUCTIVE CIRCUIT

When components such as resistors are connected to a source of e.m.f. so that an electric current is able to flow from the source through the component(s) back to the source, we call the arrangement an electrical *circuit* or, more precisely, an electrically conductive circuit. Figure 9.4 shows such a circuit, consisting of a source of e.m.f.—in this case a battery—and a single resistor. There is a number of important points to note about this simple diagram.

The source of e.m.f. is a battery and it is connected to a single resistor with wires. The circuit diagram (figure 10.4a) uses symbols

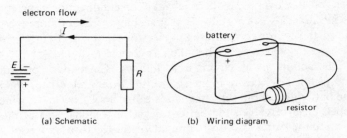

Simple Conductive Circuit

Figure 9.4

to indicate the battery—shown as E—and resistor—shown as R. This kind of diagram is called a *schematic* diagram. The other kind of diagram (figure 10.4b), which shows drawings of what the battery and resistor actually look like, is called a *wiring* diagram. Wiring diagrams show how components are placed in space relative to one another; schematic diagrams show only how components are connected together electrically and do not necessarily show where components are situated.

The circuit is one in which electric current flows in *one* direction only; such a current is called a *direct current*, abbreviated to d.c. An arrow is used to show the direction and the current symbol is shown as I. The battery, which uses chemical energy to make the current flow, has two connections or *terminals*, one shown as positive +, the other as negative −. (Batteries are discussed in more detail in chapter 12.) The direct current in the circuit is shown as flowing anticlockwise round the circuit from the positive battery terminal to the negative battery terminal. This direction of flow is called the 'conventional' direction, because it was chosen many years ago when scientists were aware that something moved in the circuit but did not know what it was. In 1895 J. J. Thomson discovered the existence of the electron, which is negative; we now know that electrons make up the current in a circuit such as this and consequently the 'conventional' direction is the opposite to that in which the charge carries actually move. We still use conventional current because many rules and aids to memory have been devised over the years which take current direction as positive to negative; also, in certain electronic components of more recent times, the charge carriers are not electrons and current flow from positive to negative may be taken as correct.

The action of the circuit is as follows. The battery provides energy for the electrons in the connecting wires and resistor and in the battery itself and sets up what is called an *electric field*, which repels electrons from the battery negative terminal and attracts them to the battery positive terminal. As the electrons move round the circuit their energy is used up in overcoming the electrical resistance of the circuit; this energy is converted mainly to heat energy as the current flows. The resistor and connecting wires become warmer as the current flows, although this is not always noticeable, since the rise in temperature may only be slight depending on the work done in moving the charge carriers. In the circuit the following relationships apply

e.m.f. = circuit resistance × current

and if the e.m.f. is taken as E volts, the circuit resistance as R ohms and the current as I amperes, then

$$E = R \times I$$

which may be taken as a mathematical statement of Ohm's law mentioned earlier.

Note that R is the *total* circuit resistance, including the resistance of connecting wires and of the resistor, when this equation is used for calculation purposes.

POTENTIAL DIFFERENCE

When electrons enter the top of resistor R in figure 9.4 they have more energy than when they leave the bottom of the resistor, the difference in energy between top and bottom being the energy used in passing through the resistor. This situation may be likened to that of a body held above ground and allowed to fall under the influence of gravity. Inside the resistor the electrons are attracted,

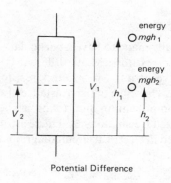

Figure 9.5

from top to bottom in figure 9.5, by the field set up by the battery; in the case of the body above ground the body is attracted from the top in figure 9.5 to the bottom by the gravitational attraction of the Earth (sometimes called a gravitational field). The potential energy of the body is higher at the top position shown than at the bottom, and if the body is allowed to fall, its potential energy is converted to kinetic energy as it does so. In the case of the electrons, the energy of the electrons at the top, as they enter the resistor, is greater than at the bottom, when they leave, and as they pass through, the energy of the electrons is converted, to heat energy on this occasion. We can say that the electrical potential energy of the electrons changes as they pass through the resistor since 'potential energy' means energy by virtue of position, and the electrons' energy does depend on their position in the circuit.

Returning to the body held above ground, if the mass is m kg then at height h_1 the potential energy of the body is mgh_1 joules, (where g is the acceleration due to gravity) and at height h_2 m above ground the potential energy is mgh_2 joules. The difference in potential energy between the two points is $(mgh_1 - mgh_2)$ joules.

Suppose the energy level per coulomb at the top of the resistor is V_1 joules/coulomb and the energy level per coulomb at the point shown in figure 9.5 is V_2 joules/coulomb; then the difference in electrical potential energy between the two points is $(V_1 - V_2)$ joules/coulomb. A joule per coulomb is of course a volt, so that we may say that the difference in electrical potential energy between the points shown is $(V_1 - V_2)$ volts.

Instead of the cumbersome phrase 'difference in electrical potential energy' we use the term *potential difference*, abbreviated to p.d.

The potential difference, or p.d., between any two points in an electrically conductive circuit is the difference between the voltage levels at the two points, the voltage levels being taken with respect to any other point. In figure 9.5 the p.d. across the whole resistor is V_1, and between the point shown and the bottom of the resistor is V_2. The p.d. between the point shown and the top of the resistor is then

$$V_1 - V_2$$

as stated above.

Returning to the circuit of figure 9.4, the energy given by the battery is E joules/coulomb and all this energy is used up as the electrons move round the circuit. Clearly the difference in energy levels per coulomb between the entry point to the circuit and the leaving point of the circuit is going to be E joules/coulomb.

For an electrically conductive circuit

total e.m.f. in the circuit = total p.d.

This point will be considered in more detail later.

Summarising, then, we may say that e.m.f. is the energy which is *given* per unit charge by a source, and p.d. is the energy *taken* from unit charge as the charge moves. Alternatively, p.d. is the energy available from unit charge because of its position relative to some fixed point, just as the potential energy of a body with respect to ground level is the energy which we can obtain from the body (usually as kinetic energy) because of its position relative to the ground.

MORE COMPLEX CIRCUITS : RESISTORS IN SERIES AND IN PARALLEL

When components are connected so that the same current flows through each—not just the same value but the same current—they are said to be connected in *series*.

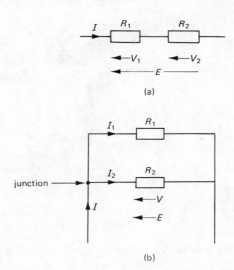

(a)

(b)

Resistors in Series and in Parallel

Figure 9.6

Figure 9.6a shows two resistors connected in series. The same current, I, flows through each resistor. The current is caused by an e.m.f., shown as E, and a p.d. exists across each resistor, its value depending on the value of the resistor since this determines how much energy is required from the charge carriers as they move through the resistor. The p.d. across resistor R_1 is shown as V_1 and across resistor R_2 is shown as V_2. The total energy per unit charge supplied by the source is E joules/coulomb, the energy per unit charge used in passing through R_1 is V_1 joules/coulomb and the energy per unit charge used in passing through R_2 is

V_2 joules/coulomb. The total energy per unit charge used in passing through the series connection of R_1 and R_2 is $V_1 + V_2$. Thus, $E = V_1 + V_2$.

For resistor R_1, voltage = current × resistance, that is

$$V_1 = I \times R_1$$

For resistor R_2

$$V_2 = I \times R_2$$

and

$$V_1 + V_2 = IR_1 + IR_2$$
$$= I(R_1 + R_2)$$

so that

$$E = I(R_1 + R_2)$$

For the circuit as a whole

total voltage (e.m.f.) = current × total resistance

and by comparison we see that the effective total resistance of this connection is the sum of the resistance of the two components connected in series.

The total resistance of components connected in series is the sum of the individual resistances.

When components are connected so that the same voltage is applied across each, they are said to be in *parallel*. The resistors shown in figure 9.6b are in parallel. Here, the e.m.f., E, is applied across R_1 and R_2 whose ends are connected together so that the same voltage exists across each. The potential difference across each is the same and is equal to the e.m.f. Each resistor carries a current determined by the value of the resistor; the higher the resistance of the resistor the lower the current, since voltage/resis-

tance is equal to current. The currents are shown as I_1 and I_2, the total current being shown as I.

Since there is no alternative path for the charge carriers to take, the number of carriers approaching the junction shown is equal to the number leaving the junction. Current means the amount of charge flowing per second, and, since each carrier has the same charge, the current depends on the number of charge carriers. Hence we can say that if the number of carriers approaching the junction equals the number of carriers leaving the junction, then the current flowing towards the junction, I, is equal to the sum of the currents I_1 and I_2 leaving the junction.

$$I = I_1 + I_2$$

For resistor R_1, voltage = current × resistance or current = voltage/resistance, thus

$$I_1 = \frac{V}{R_1}$$

For resistor R_2

$$I_2 = \frac{V}{R_2}$$

total current $= I_1 + I_2$

$$= \frac{V}{R_1} + \frac{V}{R_2}$$

$$= V\left(\frac{1}{R_1} + \frac{1}{R_2}\right)$$

For the circuit as a whole

total current $= \dfrac{\text{total voltage}}{\text{total resistance}}$

$$= \frac{V}{\text{total resistance}}$$

and by comparison we see that the total resistance of the circuit as a whole is equal to the reciprocal of $1/R_1 + 1/R_2$. Or, denoting total resistance by R

$$\frac{V}{R} = \frac{V}{R_1} + \frac{V}{R_2}$$

and

$$\frac{1}{R} = \frac{1}{R_1} + \frac{1}{R_2}$$

dividing throughout by V.

The total resistance of resistors connected in parallel is the reciprocal of the sum of the reciprocals of the individual resistances.

As is often the case, expressing a principle in words rather than symbols makes the principle appear more complex than is necessary. A useful quantity we can use here is *conductance*, which is the reciprocal of resistance. Conductance is a measure of the ease with which an electric current can be established in a circuit or component, whereas resistance is a measure of the difficulty experienced when establishing current.

$$\text{Conductance} = \frac{1}{\text{resistance}}$$
$$= \frac{\text{current}}{\text{voltage}}$$

The symbol normally used is G and the unit is the *siemens*, symbol S. (Note that the singular of the unit is siemens, not siemen.) Replacing $1/R$ by G, $1/R_1$ by G_1 and $1/R_2$ by G_2 in the equation above we have

$$G = G_1 + G_2$$

and we can say that *the total conductance of resistors connected in parallel is the sum of the individual conductances.*

The following examples should be studied carefully. They demonstrate a number of basic electrical principles.

Example 9.1

When 5 V is applied across a certain resistor, a current of 2 A flows. Calculate the resistance.
Solution

$$\text{Resistance} = \frac{\text{voltage}}{\text{current}}$$
$$= \frac{5}{2}$$
$$= 2.5 \ \Omega$$

Example 9.2

Find the resistance and conductance of a resistor if a current of 30 mA flows when a voltage of 100 V is applied across it.
Solution   A current of 30 mA (30 milliamperes) means 30/1000 amperes.

$$\text{Resistance} = \frac{\text{voltage}}{\text{current}}$$
$$= \frac{100}{30/1000}$$
$$= 100 \times \frac{1000}{30}$$
$$= 3333.3 \ \Omega$$
$$= 3.333 \ \text{k}\Omega \ \text{(kilohms)}$$
$$\text{conductance} = \frac{1}{\text{resistance}}$$
$$= \frac{1}{3333.3}$$
$$= 0.0003 \ \text{S}$$

Example 9.3

The following is a table of observations of voltages, currents and resistances. Fill in the gaps.

Voltage	Current	Resistance
10 V	3 A	?
?	20 mA	65 Ω
87 mV	?	1.2 kΩ
55 kV	90 μA	?

Line one

$$\text{resistance} = \frac{10}{3}$$
$$= 3.33 \ \Omega$$

Line two

$$\text{voltage} = \frac{20}{1000} \times 65$$
$$= 1.3 \ \text{V}$$

Line three

$$\text{current} = \frac{87/1000}{1200}$$
$$= \frac{87}{1200 \times 1000}$$
$$= 0.000\,072\,5$$
$$= 72.5 \ \mu\text{A}$$

Line four

$$\text{resistance} = \frac{55\,000}{90/1\,000\,000}$$
$$= \frac{55\,000\,000\,000}{90}$$
$$= 611\,111\,111$$
$$= 611.1 \ \text{M}\Omega$$

(In calculations involving submultiples such as this a better idea is to use powers of ten once you are able to.)

Example 9.4

Find the resistance of R in the circuit in figure 9.7.

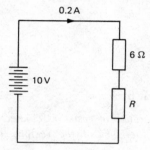

Figure 9.7

Solution

$$\text{Total resistance of circuit} = \frac{\text{e.m.f.}}{\text{current}}$$
$$= \frac{10}{0.2}$$
$$= 50 \ \Omega$$

but

$$\text{total resistance} = 6 + R$$

(resistors in series) hence

$$50 = 6 + R$$

and

$$R = 44 \ \Omega$$

Example 9.5

Calculate the value of the total resistance when three resistors of resistance 15 Ω, 27 Ω and 33 Ω are connected in parallel.
Solution   Let R be the required total resistance. Then

$$\frac{1}{R} = \frac{1}{15} + \frac{1}{27} + \frac{1}{33}$$
$$= 0.134$$
$$R = 7.46\ \Omega$$

Example 9.6

Find the value of R in the circuit shown in figure 9.8.

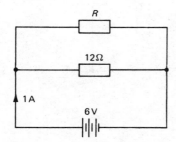

Figure 9.8

Solution

$$\text{Total circuit resistance} = \frac{\text{voltage}}{\text{current}}$$
$$= \frac{6}{1}$$
$$= 6\ \Omega$$

and

$$\frac{1}{\text{circuit resistance}} = \frac{1}{12} + \frac{1}{R}$$

that is

$$\frac{1}{6} = \frac{1}{12} + \frac{1}{R}$$
$$\frac{1}{R} = \frac{1}{6} - \frac{1}{12}$$
$$= \frac{1}{12}$$

hence

$$R = 12\ \Omega$$

Example 9.7

In the circuit shown in figure 9.9 find (a) total circuit resistance and conductance, (b) the p.d. across each resistor and (c) the value of I_1, I_2 and I.

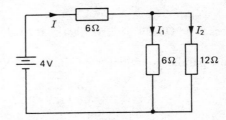

Figure 9.9

Solution   (a) The total resistance is the resistance of a 6 Ω resistor in series with the parallel combination of two resistors, one 6 Ω and one 12 Ω. Let R be the resistance of the parallel combination.

$$\frac{1}{R} = \frac{1}{6} + \frac{1}{12}$$
$$= \frac{12 + 6}{72}$$

$$= \frac{1}{4}$$

hence

$$R = 4\,\Omega$$

Hence

total resistance $= 4 + 6 = 10\,\Omega$

conductance $= \dfrac{1}{10} = 0.1$ siemens

(b) The current I is the total current, and total current from source is found from total voltage/total resistance, hence

$$I = \frac{4}{10}$$

$$= 0.4\,\text{A}$$

The p.d. across the $6\,\Omega$ resistor is $6I$, that is, $6 \times 0.4 = 2.4$ V, then

p.d. across parallel combination $=$ e.m.f. $- 2.4$ V

$$= 4 - 2.4$$

$$= 1.6\,\text{V}$$

since the total p.d. is equal to the total e.m.f.

(c) The value of I has already been found above.

$$I_1 = \frac{\text{p.d. across } 6\,\Omega}{6}$$

$$= \frac{1.6}{6}$$

$$= 0.267\,\text{A}$$

$$I_2 = I - I_1$$

$$= 0.4 - 0.267$$

$$= 0.133\,\text{A}$$

(Alternatively, $I_2 = $ (p.d. across $12\,\Omega)/12 = 1.6/12 = 0.133$ A as before.)

Example 9.8

Find the total resistance and conductance of the circuit shown in figure 9.10.

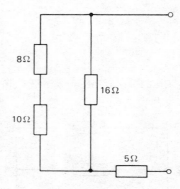

Figure 9.10

Solution $8\,\Omega$ in series with $10\,\Omega$ is equivalent to $18\,\Omega$. This $18\,\Omega$ resistance is in parallel with $16\,\Omega$ and so the equivalent resistance R is given by

$$\frac{1}{R} = \frac{1}{18} + \frac{1}{16}$$

Hence

$$R = 8.47\,\Omega$$

This $8.47\,\Omega$ resistance is in series with the $5\,\Omega$ resistor and the total equivalent resistance of the whole circuit is

$$8.47 + 5 = 13.47\,\Omega$$

$$\text{total conductance} = \frac{1}{13.47} = 0.074 \text{ siemens}$$

RESISTIVITY AND CONDUCTIVITY

In the earlier discussion concerning conductors and insulators, care was taken to mention the relative size of materials when examining their electrical resistance. It was stated that 'if two similar sized pieces of material are examined, one an insulator and one a conductor, it will be found that the conductor has a much lower electrical resistance than the insulator'. If a particular value of resistance in ohms is chosen, we can arrange for any piece of material, conductor or insulator, to have that value of resistance—provided we arrange the size of the material accordingly. For a particular resistance value the piece of insulator will be smaller than the conductor because the atomic structure of the insulator gives it a larger inherent resistance per unit size. In other words if, say, a metre cube (a cube of side length one metre) is taken of each type of material, the insulator will have a larger resistance between opposite faces than will the conductor. The value of the resistance between opposite faces of a metre cube of any material is called the *specific resistance* or *resistivity* of the material. Resistance is a general term indicating opposition to current flow and no mention of size is made. Resistivity on the other hand implies a particular size, that of a cube of side one metre. The symbol used for resistivity is ρ (pronounced 'ro'); we shall examine its units shortly; first let us consider the relationship between resistance and resistivity.

The resistance of any piece of material to current flow is directly proportional to the length of material which the current has to flow through. A simple analogy is to consider the opposition encountered when walking down a crowded street—the longer the street the more the total difficulty encountered in trying to walk down it.

Resistance is directly proportional to length.

The resistance of any piece of material is *inversely* proportional to its cross-sectional area, that is, the area of a cross-section taken at right-angles to the direction of current flow. (Returning to our busy street, we can see that the wider the street the easier it is to walk down it when it has many people in it.)

Resistance is inversely proportional to cross-sectional area.

For a cube of side one metre

$$\text{resistance} = \text{resistivity (by definition)}$$

For a cube of side l metres

$$\text{resistance} = \text{resistivity} \times l$$

since resistance is directly proportional to length, and for a piece of material of cross-sectional area a square metres (not necessarily a cube) and length l metres

$$\text{resistance} = \text{resistivity} \times \frac{l}{a}$$

since resistivity is inversely proportional to area. The general relationship is then

$$\text{resistance} = \frac{\text{resistivity} \times \text{length}}{\text{cross-sectional area}}$$

or

$$\text{resistivity} = \frac{\text{resistance} \times \text{cross-sectional area}}{\text{length}}$$

and we see that the units of the right-hand side of the equation are ohm-metre^2 per metre or (cancelling 'metres' in numerator and denominator) the unit of resistivity is ohm-metres. Resistance is measured in ohms alone, but as we can see, the size of the material comes into the resistivity unit.

Conductivity is the conductance between opposite faces of a cube of side one metre and, as such, is the reciprocal of resistivity. Its symbol is σ (sigma). Since the units of resistivity are ohm-metre^2/metre, the units of conductivity are metre/ohm-metre^2, that is

$$\frac{\text{metre}}{\text{ohm} \times \text{metre}^2}$$

but 1/ohm is called a siemens so the units of conductivity are

$$\frac{\text{metre-siemens}}{\text{metre}^2}$$

or siemens/metre.

Example 9.9

Calculate the resistance of a piece of copper of length 3.4 m and cross-sectional area 0.53 m^2 if the resistivity of the material is 1.56 $\times$ 10^{-8} ohm-metre.
Solution

$$\begin{aligned}
\text{Resistance} &= \frac{\text{resistivity} \times \text{length}}{\text{cross-sectional area}} \\
&= \frac{1.56}{10^8} \times \frac{3.4}{0.53}
\end{aligned}$$

(note that 10^{-8} = 1/10^8)

$$= \frac{1}{10^7} \, \Omega$$

Example 9.10

Calculate the resistance of a piece of porcelain with the same dimensions as the copper in example 9.9, if the resistivity of porcelain is 10^{12} ohm-metre.
Solution

$$\text{Resistance} = \frac{3.4}{0.53} \times 10^{12}$$

$$= 6.41 \times 10^{12} \, \Omega$$

The difference in resistance of two equal sized pieces of materials, one a conductor and the other an insulator, can readily be seen.

Example 9.11

The resistance of a certain piece of material is 80 Ω. Calculate the resistance of a piece of the same material with a length equal to twice that of the first and a cross-sectional area equal to one-half that of the first.
Solution   Since resistance is directly proportional to length, if the length is *doubled*, the resistance is *doubled*. Since resistance is inversely proportional to area, if the area is *halved* the resistance is *doubled*. The resistance is thus increased *fourfold* on the original value, thus

$$\text{resistance} = 4 \times 80 = 320 \, \Omega$$

Example 9.12

A piece of material has a cross-sectional area of 0.018 m^2, a length of 0.4 m and its resistance is 23.8 Ω. Calculate the resistance of a piece of the same material of length 0.75 m and cross-sectional area 0.04 m^2.
Solution   The problem can be solved by ratio and the value of resistivity need not be known. The length is increased in the ratio 0.75/0.4. Thus the resistance is increased in the same ratio, 0.75/0.4.

The area is increased in the ratio 0.04/0.018, thus the resistance is *reduced* in the ratio 0.018/0.04.

The new resistance will then be (0.75/0.4) $\times$ (0.018/0.04) times the old, that is

$$\begin{aligned}
\text{new resistance} &= \frac{0.75}{0.4} \times \frac{0.018}{0.04} \times 23.8 \\
&= 20.08 \, \Omega
\end{aligned}$$

EFFECT OF TEMPERATURE ON RESISTANCE

The resistance of any material is dependent on the dimensions of the material and on its resistivity, as discussed above. When a material is subjected to temperature change all these characteristics may change. Physical dimensions change with temperature (for example, metals normally expand) and since changing temperature indicates a change in heat energy of a piece of material, the total internal energy of the material changes. This affects movement of charge carriers through the material, the usual change being that as temperature *increases* the number of charge carriers passing any particular point at any one time *reduces* (due to an increasing number of collisions between them) and thus the current *reduces*. Electrical resistance usually (but not always—it depends on the material) rises with temperature.

Temperature Coefficient of Resistance

For most materials over a small temperature range the resistance varies linearly with temperature. Thus if resistance and temperature of a material are noted at various points as the temperature is increased, and a graph plotting resistance against temperature is drawn, the general shape is as that shown in figure 9.11. The temperature scale here starts at $0\,^\circ$C and R_0 is the resistance at $0\,^\circ$C. Over a temperature interval t, as shown, the resistance increases by the amount indicated and the resistance–temperature graph is a straight line.

It is found that the change in resistance is *directly proportional to the temperature change* and *directly proportional to the resistance at $0\,^\circ$C*. Thus we can say that

$$\text{resistance } change = \text{constant} \times \text{temperature change} \times \text{resistance at } 0\,^\circ\text{C}$$

The constant is called the *temperature coefficient of resistance* and the usual symbol is α (alpha). Thus the change in resistance $= \alpha \times R_0 \times t$ for the figure shown, and the new total resistance at $t\,^\circ$C, R_t say, is given by

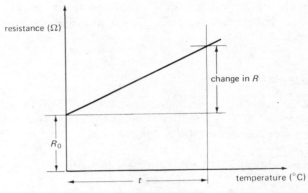

resistance (Ω)

change in R

R_0

t

temperature $(^\circ$C)

Resistance–Temperature Graph

Figure 9.11

$$R_t = R_0 + \alpha R_0\, t$$

or

$$R_t = R_0(1 + \alpha t)$$

Or, in words, the resistance at any temperature is equal to the resistance at $0\,^\circ$C multiplied by one plus the product of the temperature coefficient of resistance and the temperature change.

If the resistance at two temperatures, t_1 and t_2 say, is R_1 and R_2 respectively, then since

$$R_1 = R_0(1 + \alpha t_1)$$

and

$$R_2 = R_0(1 + \alpha t_2)$$

then

$$\frac{R_2}{R_1} = \frac{1 + \alpha t_2}{1 + \alpha t_1}$$

Thus we have a method of finding one of these quantities, knowing the others but not knowing R_0. This is shown in the following examples.

Example 9.13

The resistance of a copper conductor at $0\,^{\circ}$C is $80\ \Omega$. Calculate the resistance at $20\,^{\circ}$C given that the temperature coefficient of resistance of copper is $0.004\,26$ ohm/ohm $^{\circ}$C.

Solution

$$\text{Resistance at } 20\,^{\circ}\text{C} = 80\,(1 + 0.004\,26 \times 20)$$
$$= 86.816\ \Omega$$

Note the units of the temperature coefficient of resistance. Since the change in resistance (ohms) is equal to the temperature coefficient $\times$ temperature ($^{\circ}$C) $\times$ resistance at $0\,^{\circ}$C (ohms) then, dimensionally

$$\text{ohms} = \text{temperature coefficient} \times {^{\circ}}\text{C} \times \text{ohms}$$

and temperature coefficient units are ohms/ohms $^{\circ}$C (or per $^{\circ}$C, but the ohms are usually left in both numerator and denominator).

Example 9.14

The resistance of a certain piece of material is $30\ \Omega$ at $10\,^{\circ}$C and $33.2\ \Omega$ at $15\,^{\circ}$C. Calculate the temperature coefficient of resistance.

Solution Let the temperature coefficient of resistance be α and the material resistance at $0\,^{\circ}$C be R_0 (not actually required—it will cancel), then

$$30 = R_0(1 + 10\alpha)$$
$$33.2 = R_0(1 + 15\alpha)$$

Dividing

$$\frac{33.2}{30} = \frac{1 + 15\alpha}{1 + 10\alpha}$$

Hence

$$33.2\,(1 + 10\alpha) = 30\,(1 + 15\alpha)$$
$$33.2 + 332 = 30 + 450\alpha$$
$$3.2 = 118\alpha$$

and

$$\alpha = \frac{3.2}{118}$$
$$= 0.027\ \Omega/\Omega\,^{\circ}\text{C}$$

Example 9.15

The resistance of a piece of material at $15\,^{\circ}$C is $23\ \Omega$. Its temperature coefficient of resistance is $0.007\ \Omega/\Omega\,^{\circ}$C. Calculate its resistance at $20\,^{\circ}$C.

Solution Let the resistance at $20\,^{\circ}$C be R_{20}, then

$$\frac{R_{20}}{23} = \frac{1 + 0.007 \times 20}{1 + 0.007 \times 15}$$
$$= 1.0316$$

and

$$R_{20} = 1.0316 \times 23$$
$$= 23.728\ \Omega$$

POWER IN D.C. CIRCUITS

The unit of voltage is the joule per coulomb. The unit of current is the coulomb per second. Thus if voltage is multiplied by current we have

$$\text{voltage} \times \text{current} = \frac{\text{joule}}{\text{coulomb}} \times \frac{\text{coulomb}}{\text{second}}$$
$$= \frac{\text{joule}}{\text{second}}$$

and the joule per second (or watt) is the rate of using energy or power.

Thus to obtain the power in a circuit or component carrying direct current we multiply the voltage (p.d.) by the current. Denoting power by P, voltage by V and current by I we have

$$P = VI$$

and if the circuit or component resistance is denoted by R, since voltage is equal to current × resistance ($V = IR$) we have

$$P = IR \times I$$

(replacing V by IR). Thus

$$P = I^2 R$$

and

$$P = V\left(\frac{V}{R}\right)$$

(replacing I by V/R). Thus

$$P = \frac{V^2}{R}$$

or, in words

$$\begin{aligned} \text{power} &= \text{voltage} \times \text{current} \\ &= \text{current}^2 \times \text{resistance} \\ &= \frac{\text{voltage}^2}{\text{resistance}} \end{aligned}$$

The unit of power is the watt, with its multiples the kilowatt, megawatt and so on. This is used to derive the commercial unit of electrical energy (the 'unit' we pay for when paying electricity bills). If one kilowatt is used for one hour the *energy* consumed (not power) is called one kilowatt-hour, abbreviated k Wh. The follow-ing examples illustrate power and energy calculations.

Example 9.16

Two resistors of resistance 25 Ω and 50 Ω are connected in series across a 250 V d.c. supply. Calculate the power consumed by the total circuit and by each resistor.
Solution

$$\text{Total resistance} = 25 + 50$$

(resistors in series)

$$= 75 \ \Omega$$

$$\begin{aligned} \text{Total power} &= \frac{\text{voltage}^2}{\text{resistance}} \\ &= \frac{250^2}{75} \\ &= 833.33 \ \text{W} \end{aligned}$$

The circuit current (in each resistor since they are in series) is given by applied voltage/total resistance, thus

$$\begin{aligned} \text{circuit current} &= \frac{250}{75} \\ &= 3.33 \ \text{A} \end{aligned}$$

Power in each resistor is found from power = current² × resis-tance, thus

$$\begin{aligned} \text{power in 25 } \Omega \text{ resistor} &= 3.33^2 \times 25 \\ &= 277.22 \ \text{W} \\ \text{power in 50 } \Omega \text{ resistor} &= 3.33^2 \times 50 \\ &= 554.44 \ \text{W} \end{aligned}$$

(The sum of these does not exactly equal 833.33, since only two decimal places are taken.)

Example 9.17

Determine the minimum power rating of a 33 kΩ resistor to be

connected across a 24 V d.c. supply.
Solution

$$\text{Power} = \frac{\text{voltage}^2}{\text{resistance}}$$
$$= \frac{24^2}{33\,000}$$
$$= 0.017 \text{ W}$$

The smallest power rating normally obtainable is one-eighth watts; this is more than ample for this resistor.

The power rating of resistors should always be checked before insertion into a circuit. If the power absorbed exceeds the power rating, the resistor will burn out.

Example 9.18

Calculate the cost of running the following equipment for the times stated

1 1 kW heater for 5 hours
4 100 W lamps for 15 hours
1 245 W TV receiver for 4 hours
1 750 W electric iron for 45 minutes
1 15 W soldering iron for $2\frac{1}{2}$ hours

The cost per unit may be taken as 2.85p.
Solution   In all such calculations the following equation is used

$$\text{cost} = \text{number of items} \times \text{power rating in kW}$$
$$\times \text{hours of use} \times \text{cost/unit}$$

For the heater

$$\text{cost} = 1 \times 1 \times 5 \times 2.85$$
$$= 15.25\text{p}$$

For the lamps

$$\text{cost} = 4 \times 0.1 \times 15 \times 2.85$$
$$= 17.1\text{p}$$

For the TV receiver

$$\text{cost} = 1 \times 0.245 \times 4 \times 2.85$$
$$= 2.793\text{p}$$

For the electric iron

$$\text{cost} = 1 \times 0.75 \times \frac{45}{60} \times 2.85$$
$$= 1.603\text{p}$$

For the soldering iron

$$\text{cost} = 1 \times 0.015 \times 2.5 \times 2.85$$
$$= 0.107\text{p}$$

The total cost is the sum of all these, which is 35.8 pence.

ASSESSMENT EXERCISES

Long Answer

9.1 Fill in the blanks in the following table.

Voltage	Current	Resistance
10 V	3 A	
20 mV		1 kΩ
	10 mA	6.8 kΩ
0.7 kV		100 Ω
25 mV	75 μA	
	0.02 A	680 Ω

9.2 The current flowing in a circuit composed of three equal resistors in series is 150 mA when 25 V are applied. Calculate the value of each resistance.

9.3 Three resistors, 1.2 kΩ, 5.6 kΩ and 10 kΩ are connected in parallel across a 100-V d.c. supply. Find the current in each resistor.

9.4. A battery of e.m.f. 9 V is connected across a resistor of resistance 100 Ω and the current flowing is found to be 82 mA. Find the value of the resistance of the leads and battery.

9.5 Two 8.2-kΩ resistors connected in parallel are then connected in series across a d.c. supply. The current flowing is 10 mA. Find the p.d. across the circuit.

9.6 A resistor R is connected in parallel with a 100-Ω resistor; the combination is then connected in series with a 50-Ω resistor. When 100 V d.c. are applied to the circuit as a whole a current of 1 A flows. Find the value of resistance of resistor R.

9.7 Three resistors of resistance 100 Ω, 150 Ω and 470 Ω are connected in series; the combination is then connected in parallel with a 1-kΩ resistor. Find the conductance of the circuit.

9.8 Calculate the resistance of a piece of copper of length 5 m and cross-sectional area 0.7 m^2. (Resistivity of copper is 1.56 $\times 10^{-8}$ Ω m)

9.9 A bar of ceramic material has a cross-sectional area of 0.002 m^2 and a length of 1.3 m. The resistivity of the material is 10^{12} Ω m. Determine the resistance of the bar.

9.10 A piece of material has the dimensions cross-sectional area 0.02 m^2, length 0.3 m and has a resistance of 45 Ω. Calculate the resistance of a piece of the same material of area 0.01 m^2 and length 0.37 m.

9.11 The resistance of a copper conductor at 0 °C is 120 Ω. Determine the resistance at 15 °C. (Temperature coefficient of

resistance of copper is 0.004 26 Ω/Ω °C)

9.12 The resistance of a certain piece of material is 80 Ω at 15 °C and 80.7 Ω at 20 °C. Calculate the temperature coefficient of resistance.

9.13 Two resistors of resistance 1.5 kΩ and 3.3 kΩ respectively are connected (a) in series (b) in parallel across a 50-V d.c. supply. Calculate the power in each resistor and in the total circuit in each case.

9.14 A circuit consisting of a 470-Ω resistor in series with a 680-Ω resistor is connected across a 100-V d.c. supply. Determine the minimum power rating of each resistor.

9.15 A resistor of resistance 820 Ω and rated 0.5 W is connected in series with a resistor of resistance 220 Ω rated 0.25 W and the circuit is connected across a d.c. supply. Determine the maximum voltage that the supply may have without damage occuring to the resistors, and the power dissipated in each resistor when the supply has this value of voltage.

9.16 Find the maximum current that can be passed through the following resistors (the details given are resistance and power rating): 220 Ω, 0.5 W; 560 Ω, 0.25 W; 1 kΩ, 1 W; 470 Ω, 0.125 W; 1200 Ω, 2 W.

9.17 Calculate the cost of running the following equipment: one 1-kW heater for 6 h, one 330-W infrared heat/light for 10 h, six 150-W lamps for 2 h, one 750-W electric iron for 30 min; the cost per unit is 2.67p.

Short Answer

9.18 State the units of e.m.f., p.d., resistance, conductance and electric current.

9.19 Calculate the current flowing in a 1.2-kΩ resistor connected across a 10-V d.c. supply.

9.20 Calculate the voltage across a 4.7-kΩ resistor in which a current of 22 mA flows.

9.21 Calculate the resistance of a resistor in which a current of 42 mA flows when it is connected across a 110-V d.c. supply.

9.22 Calculate the equivalent resistance of three resistors of value 1.2 kΩ, 3.3 kΩ and 4.7 kΩ connected in series.

9.23 Calculate the equivalent resistance of three resistors of value 4.7 kΩ, 6.8 kΩ and 10 kΩ connected in parallel.

9.24 Calculate the power dissipated in a 4.7-kΩ resistor connected across a 10-V d.c. supply.

9.25 Calculate the power dissipated in a 6.8-kΩ resistor carrying a current of 10 mA.

9.26 Calculate the power dissipated in a circuit which when connected to a 9-V d.c. supply takes a current of 1.2 A.

9.27 Determine the conductance of a resistor of value 15 kΩ.

9.28 Define resistivity.

9.29 Calculate the resistance of a piece of material of length 1.7 m, cross-sectional area 0.01 m^2 and which has a resistivity of 0.00426 Ω m.

9.30 A resistor has a resistance of 10 Ω at 0 °C. The temperature coefficient of resistance is 0.002 Ω/Ω °C. Find the resistance of the component at 25 °C.

9.31 Determine the total number of commercial energy units required by a piece of electrical apparatus rated at 1.75 kW when it is run for 2 h 25 min.

9.32 Define the commercial energy unit.

Multiple Choice

9.33 The unit of electric charge is the
  A. coulomb   B. volt   C. ohm   D. ampere

9.34 The unit of electric current is the
  A. coulomb/second   B. volt   C. ohm   D. coulomb

9.35 The unit of electrical resistance is the
  A. siemens   B. ampere   C. ohm   D. volt

9.36 The total equivalent resistance of two resistors of resistance R_1 and R_2 in series is equal to

  A. $\dfrac{1}{R_1} + \dfrac{1}{R_2}$ B. $\dfrac{R_1 R_2}{R_1 + R_2}$ C. $R_1 + R_2$ D. $\dfrac{1}{R_1 + R_2}$

9.37 The total equivalent resistance of two resistors of resistance R_1 and R_2 in parallel is equal to
  A. $\dfrac{1}{R_1} + \dfrac{1}{R_2}$ B. $\dfrac{R_1 R_2}{R_1 + R_2}$ C. $R_1 + R_2$ D. $\dfrac{1}{R_1 + R_2}$

9.38 The resistance of a piece of material of resistivity ρ, length l and cross-sectional area a (all in SI units) is given by
  A. $\dfrac{l}{\rho a}$ B. $\dfrac{\rho a}{l}$ C. $\dfrac{\rho l}{a}$ D. $\dfrac{a}{\rho l}$

9.39 A piece of material has an electrical resistance of 10 Ω. If its length is doubled and its cross-sectional area halved the new resistance is
  A. 10 Ω B. 40 Ω C. 2.5 Ω D. incalculable without more information

9.40 The resistance of a piece of material at a temperature t °C in terms of its resistance at 0 °C, R_0 Ω, and the temperature coefficient of resistance α is given by
  A. $1 + \alpha t$ B. $R_0 + \alpha t$ C. $R_0 + \alpha R_0 t$ D. $\alpha R_0 t$

9.41 The resistance R_2 of a material at any temperature t_2 in terms of its resistance R_1 at any other temperature t_1 and the temperature coefficient of resistance α is given by

A. $\dfrac{R_2}{R_1} = \dfrac{1 + \alpha t_1}{1 + \alpha t_2}$ B. $\dfrac{R_2}{R_1} = \dfrac{1 + \alpha t_2}{1 + \alpha t_1}$

C. $R_2 = R_1(1 + \alpha t_1)(1 + \alpha t_2)$ D. $R_2 = R_1 + \alpha t_2 - \alpha t_1$

9.42 Power in a d.c. circuit in terms of voltage V, current I and resistance R is given by

A. IR B. V/R C. VI D. $\dfrac{VI}{R}$

9.43 The unit of energy used in cost calculations is the
A. watt   B. kilowatt   C. kilowatt-hour   D. joule

9.44 The minimum power rating of a 10-Ω resistor connected across a 100-V supply is
A. 10 W   B. 100 W   C. 1 W   D. 1 kW

10 Electromagnetism

OBJECTIVES

All the objectives should be understood to be prefixed by the words
'The expected learning outcome is that the student . . .'

D13 Describes the magnetic field concept and the relationships
between magnetic fields and electric current
13.1 States that a magnet experiences a force when in a
magnetic field.
13.2 States that a current-carrying conductor produces a
magnetic field.
13.3 Describes the type of magnetic field pattern produced by
(a) a bar magnet (b) a solenoid.
13.4 States that a current-carrying conductor experiences a
force when in a magnetic field.
13.5 Explains, in terms of 13.4, the basic operation of a
moving-coil meter.
13.6 Explains, in terms of 13.4, the basic operation of a simple
d.c. motor.
13.7 Describes what happens when a permanent magnet is
moved in a coil of wire connected to a galvonometer.
13.8 Explains in terms of 13.7, the basic operation of an a.c.
generator.

The effects of magnetism, or, to be more precise, electromagnetism, have been known for a long time. In both early Chinese and early Greek civilisations it was known that pieces of certain materials, when allowed to hang freely, always pointed in a certain direction; the fact was used as an aid to the navigation of ships. Electricity and magnetism are in fact related, magnetism being caused by the movement of electric charge (as we shall discuss shortly) but the connection between the two was not suggested before the seventeenth century and was not completely verified until the twentieth century.

Magnetism is the property of setting up a force which attracts certain materials. Not all materials are affected by a magnetic force; those which are affected are called *ferromagnetic* materials, the most common being iron, nickel and cobalt. Materials containing these elements also exhibit magnetic properties. Materials containing iron are called *ferrous* materials, the stem of the word ferromagnetic and ferrous being the same (it is derived from the Latin word for iron). The material used by the Greeks as an early form of ship's compass, called *lodestone*, is now known to be a ferrous material. The magnetic force which causes the lodestone always to set in the same direction is exerted by the Earth itself.

Anything which sets up a magnetic force is called a *magnet*. A piece of ferromagnetic material is not necessarily a magnet, that is, it does not necessarily *exert* a magnetic force, although it will always react to a magnetic force exerted by another magnet. Whether or not a ferromagnetic material behaves as a magnet depends on the material's state of *magnetisation*. This will be further discussed later.

If a piece of ferromagnetic material which is a magnet (usually called a bar magnet if that is its shape) is freely suspended it will always set itself in a certain direction, one end pointing north, the other south. The end pointing north is called the *north-seeking* or *north pole* of the magnet, the other end being the *south-seeking* or *south pole* of the magnet.

MAGNETIC FIELDS : MAGNETIC FLUX

The area surrounding a magnet in which magnetic forces can be felt is called a *magnetic field*. One of the most famous nineteenth century scientists, Michael Faraday, suggested that to represent a magnetic field in diagram form we should use lines to show the direction of action of the force set up due to magnetism. These lines he called *lines of force, lines of magnetic flux* or just magnetic flux. Originally the number of lines drawn showed the relative strength of the force within the magnetic field; nowadays although we still use the idea of magnetic flux, the strength of the magnetic field is measured differently.

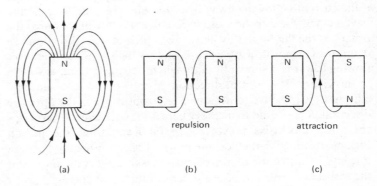

Figure 10.1

Lines of force are drawn conventionally from the north pole of the magnet to the south pole and the *field pattern* of a simple bar magnet is shown in figure 10.1a. This pattern can easily be demonstrated by placing a sheet of paper over a bar magnet and gently shaking iron filings over the paper. The filings tend to behave as very tiny magnets and set themselves in appropriate directions in the magnetic field to form a pattern of lines like that shown in figure 10.1a.

If two bar magnets are placed side by side there is an interaction between their respective fields and a force is set up between the magnets. If like poles are side by side (north next to north and south next to south) the force is one of repulsion (see figure 10.1b). If unlike poles are adjacent the force is one of attraction (see figure 10.1c). In general, when two fields interact, like fields (lines of force in the same direction) repel and unlike fields (lines of force in opposite directions) attract. The force set up between magnetic fields is used in the electric motor to produce motion.

The Magnetic Field Set Up by a Conductor

When an electric current flows through a conductor a magnetic field is set up around the conductor. The situation is shown in figure 10.2a. The lines of flux are circular and surround the conductor as shown in the figure. Looking at the conductor end on, if the current is flowing away from the observer, the lines of flux have a clockwise direction and if the current is flowing towards the observer the lines of flux have an anticlockwise direction. In diagrams current flowing away is shown by a cross and current flowing towards is shown by a dot when viewing end on, as shown in this figure. The current direction given is that of conventional current flow (positive to negative); electrons move in the opposite direction. A useful aid to memory here is the corkscrew—turning a corkscrew clockwise moves the point away from the person turning, turning it anticlockwise moves it back towards the person turning. The direction of turn corresponds to the flux direction and the respective direction of movement of the point corresponds to the direction of current flow producing the flux.

The strength of the magnetic field produced by an electric current flowing in a conductor may be considerably increased by winding the conductor in the form of a coil as shown in figures 10.2b and c. Figure 10.2b shows a cross-section of the coil in a plane through the coil centre parallel to the coil axis (perpendicular to the plane in which any one turn lies). The flux produced by each conductor *adds* to the flux produced by its neighbour giving a resultant increased flux as shown. Figure 10.2c shows a useful way of remembering directions in this case : looking end on at the coil, when current flows clockwise the flux direction is away from the observer and when current flows anticlockwise the flux direction is towards the observer. When the flux direction is away from the observer into the coil end, then that end behaves as the south pole of a magnet, the other end behaving as the north pole. When arrowheads are placed on the letters S and N as shown and the letters are drawn within a circle as shown, the

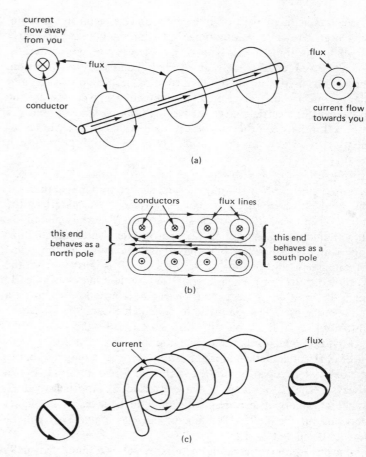

Figure 10.2

arrows show the direction of current flow in the coil. Alternatively, flux and current directions of figure 10.2a are current and flux directions respectively in figure 10.2c.

The Process of Magnetisation

As stated above, when an electric current flows a magnetic field is set up in the vicinity of the current, that is, a magnetic force is established which will have an effect on ferromagnetic materials nearby. Basic atomic theory indicates that all materials are made up of atoms, which in turn consist of a nucleus orbited by electrons. Now an electric current is a movement of electrically charged particles and thus the electrons orbiting within an atom constitute an electric current. A number of orbits produces the equivalent of a number of minute currents and each current sets up a magnetic field. In the majority of materials the small magnetic fields within the material act in various directions such that the resultant field is zero. In ferromagnetic materials, however, it is believed that some of the internal fields act together to produce a number of what are called 'magnetic domains'. These domains may be considered to behave like a large number of small bar magnets. In an un-magnetised ferromagnetic material the domains are situated at random throughout the material and no resultant field is produced.

This is illustrated in diagram form in figure 10.3 which shows the domains as arrows pointing in many different directions (figure 10.3a). If the material is placed in an external field the domains begin to line up (figure 10.3b) and a small resultant field is set up. The material is now partially magnetised and is capable of exerting its own field. Eventually as the external field is increased in strength all the domains are aligned (figure 10.3c) and the material is fully magnetised.

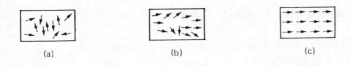

(a) (b) (c)

Figure 10.3

If the external field is removed the domains may return to the state shown in figure 10.3b, that is, the material is still partially magnetised, or may remain fully magnetised, depending on the material. Soft iron, for example, loses most of the magnetism when the field is removed but steel retains most of its magnetism. Steel and metals with similar properties (steel with nickel or certain alloys) are used for making permanent magnets. A permanent magnet retains its magnetism and is capable of setting up its own field if the magnetising field is taken away. A permanent magnet may then be used to make other magnets (the usual method uses a coil and electric current to set up the external field). Non-magnetic materials which do not have these domains are not affected by external fields at all.

Magnetic Circuit Quantities

An electrically conductive circuit is a collection of pieces of material joined together such that a source of voltage (an electromotive force or e.m.f.) causes an electric current to flow through the materials in a continuous path. The opposition to current flow, measured as voltage/current, is called electrical resistance and the reciprocal of resistance, which is a measure of the support offered by the circuit to the current, is called conductance. Many electromagnetic machines and other devices (for example, relays and circuit breakers) may be considered to be or to have a *magnetic circuit* and it is a useful exercise to compare quantities in the conductive circuit with those in the magnetic circuit.

Figure 10.4a shows a simple magnetic circuit. It consists of a ferromagnetic material in the form of a square with the centre missing, so that it is made up of four pieces or *limbs*. The left-hand limb carries a coil of wire through which an electric current flows. (The coil is part of a separate electrically conductive circuit.) Using the theory already given, lines of force or flux may be drawn through the coil, the top end of the coil behaving like the north pole of a bar magnet. Because of the nature of the circuit—it is made up of ferromagnetic material—the magnetic field is stronger within the circuit than outside it. Thus flux lines need only be drawn within the limbs as shown. The quantities in this magnetic circuit comparable to e.m.f. and current in the conductive circuit are the quantity which causes the field to be set up and some quantity which represents the field itself.

What are these quantities? First of all let us consider what sets up the magnetic field contained within the limbs. Clearly the electric

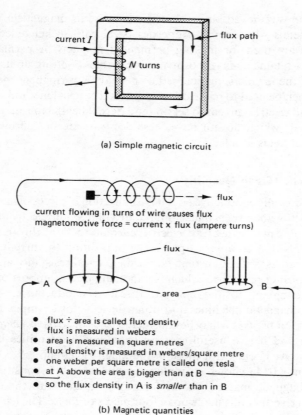

(a) Simple magnetic circuit

current flowing in turns of wire causes flux
magnetomotive force = current x flux (ampere turns)

- flux ÷ area is called flux density
- flux is measured in webers
- area is measured in square metres
- flux density is measured in webers/square metre
- one weber per square metre is called one tesla
- at A above the area is bigger than at B
- so the flux density in A is *smaller* than in B

(b) Magnetic quantities

Figure 10.4

current in the coil helps to establish the field, since without it the only magnetism is that of the material making up the limbs, if this material has magnetism (Although it is ferromagnetic it may be in a state of non-magnetisation as explained earlier.) However, it is not only the current which matters, it is also the coil, or more specifically, the number of turns on the coil. The more turns there are for any particular current value the stronger is the resulting magnetic field.

The name given to the quantity which sets up a magnetic field is *magnetomotive force*, abbreviated to m.m.f. As with the term electromotive force, the name is unfortunate since the quantity is *not* a force, although, in this case, it is responsible for the setting up of conditions in which a force may be experienced. In the SI System of units the quantity m.m.f. is measured by multiplying together the current and the number of turns; the unit is therefore the *ampere-turn*. Thus a 100-turn coil carrying a current of 0.5 A, for example, has an m.m.f. of 100×0.5 or 50 ampere-turns.

To find a magnetic quantity comparable to electric current we take the idea of flux one stage further than being merely a means of diagrammatically representing magnetic fields, and imagine the flux as being set up by the m.m.f. Although, strictly speaking, flux is not a physical quantity like electric current, it is nevertheless a very useful concept which we can use to compare magnetic fields. Flux, like current, may be considered to be the 'effect' in the circuit due to the appropriate 'cause' (m.m.f. or e.m.f. respectively). Flux is given a unit (we can measure it in terms of its effects), the unit being defined in terms of other units (discussed later). The special name given to the unit of flux is the *weber*, symbol Wb.

Thus, summarising these ideas, magnetomotive force causes magnetic flux in a similar way to electromotive force causing electric current. M.M.F. is measured in ampere-turns and magnetic flux in webers. The method of defining a weber (to give an idea of the size of the unit) is discussed later in the chapter.

Magnetic Flux Density

If we assume that magnetic flux is a measurable quantity we can go one stage further in our efforts to compare magnetic fields and consider how this flux is distributed. If the same flux is distributed in two pieces of similar material, one being of larger cross-section than the other, it seems clear that, speaking loosely, the magnetic effect in the smaller piece is greater than in the larger piece, since the same flux is concentrated in a smaller area. This leads us to the idea of *flux density*, which is the magnetic flux (webers) divided by the cross-sectional area through which the flux acts (area in square metres). The unit of flux density is thus the weber per square metre, one weber per square metre being given the special name *tesla*, symbol T (see figure 10.4b).

Force Exerted on a Conductor Placed in a Magnetic Field

Earlier it was stated that in general when magnetic fields are close to one another, a force is set up between them, either of attraction or of repulsion (figure 10.1). Now a conductor carrying an electric current has its own magnetic field, so if it is placed in an external field, a force is exerted on the conductor and, if it is able to do so, it will move. The resulting field pattern will be considered later. For the moment consider the size of this force and what it depends on. Firstly, it will depend on the flux density of the external field — the flux density rather than the flux alone because the denser the field the greater will be the force. Secondly, it will depend on the magnitude of the current carried by the conductor—because the field set up by the conductor is determined in size by the value of the current causing it and the force between fields depends in turn on the relative sizes of the fields concerned. Finally, the force depends on the length of the conductor: the longer the conductor the greater the force because more of the conductor field reacts with the applied external field.

If SI units are used the force on a conductor placed in a magnetic field is given by

force in newtons = flux density in teslas × length of
conductor in metres
× current in amperes

or, using symbols

$$F = BlI$$

where F is the force, B the flux density, l the length and I the current.

Example 10.1

Calculate the current flowing in a wire placed in a magnetic field of flux density 0.1 T if the force per metre length of the conductor is 0.5 N.
Solution

$$\text{Force} = \text{flux density} \times \text{length} \times \text{current}$$

so that

$$\text{current} = \frac{\text{force}}{\text{flux density} \times \text{length}}$$

and, substituting values

$$\text{current} = \frac{0.5}{0.1 \times 1}$$
$$= 5 \text{ A}$$

Example 10.2

A conductor of length 0.2 m carrying a current of 50 A is placed in a magnetic field of flux density 0.6 T. Calculate the force on the conductor assuming it is at right-angles to the field.
Solution

$$\text{Force} = 0.6 \times 0.2 \times 50$$
$$= 6 \text{ N}$$

Note that for the above formula to be correct the conductor is at right-angles to the field of which the flux density is used in the equation. If the conductor is at any other angle, then the flux density is that of the field *component* which is at right-angles to the conductor. (A magnetic field may be considered to have components in much the same way as a force or other vector quantity.)

The field pattern when a current-carrying conductor is placed in a magnetic field is illustrated in figure 10.5. The resultant field due to like fields repelling and unlike fields attracting is shown in the figure. It is useful to regard lines of force as being 'elastic' in their nature in that they attempt to take up the shortest length. If the lines were elastic, a force would be exerted in the direction shown,

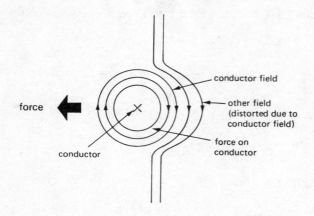

Figure 10.5

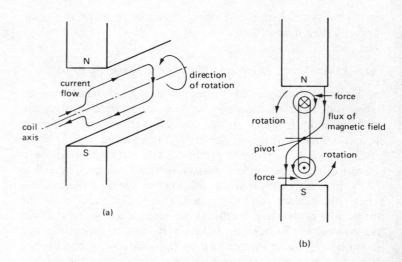

(a)

(b)

pushing the conductor to the left. The principle is used in electric motors and in indicating instruments using moving coils.

The Motor Principle

If a coil is pivoted about its axis so that it is free to rotate, and a magnetic field is set up so that the coil conductors are at right-angles to the field flux as shown in figure 10.6a, a force is exerted on both conductors of the coil as shown in figure 10.6b. The coil will tend to rotate about its axis. This is the basic principle of motors and moving-coil indicating instruments. If the principle is being used in a motor some means must be found of allowing the electric current to flow into and out of the coil as it rotates and the current must always flow in the appropriate directions shown in the figure, that is, into the paper in the top conductor and out of the paper in the bottom conductor in figure 10.6b. The simplest way of doing this is to use a *commutator*, as shown in figure 10.6c. The positive side of the supply is connected to the top brush and the negative side to the bottom brush; the brushes rub against the two segments connected to the conductors in turn as the coil rotates. In this way current always flows into the top conductor and out of the bottom conductor. In practice the main field is supplied by permanent

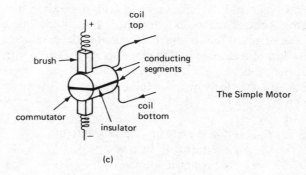

The Simple Motor

(c)

Figure 10.6

magnets only in small machines and in permanent-magnet–moving-coil meters (PMMC); in larger machines and in *dynamometer*-type meters the main field is supplied by a coil.

Some Other Applications of Electromagnetism

Figure 10.7 shows some applications of electromagnetism using a

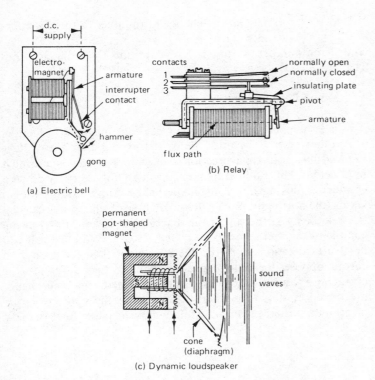

(a) Electric bell

(b) Relay

(c) Dynamic loudspeaker

Figure 10.7

'normally open' arrangement the side arm pushes the contact arms together, in the 'normally closed' arrangement the side arm goes through a hole in one of the contact arms and the contact arms are pushed apart when the relay operates. The flux path is shown by a dotted line. The relay coil may be operated by a very low voltage and the switches used to operate high-voltage circuits; thus a high-voltage circuit may be operated quite safely and without danger to the operator.

Figure 10.7c shows a moving-coil loudspeaker which changes *alternating current signals* into sound. An alternating current signal is an electric current which flows alternately backwards and forwards through the coil, the change in current magnitude being used to carry information. For example an alternating current signal is obtained from a *microphone* which turns sound into electrical impulses. The loudspeaker carries out the opposite process to the microphone. The speaker coil is free to move along the middle part of the E-shaped core. The reaction between the fixed magnets of the core and the fluctuating field of the coil causes a varying force to be set up and this in turn moves the cone in sympathy with the force and thus with the signal causing the force.

In all these applications a suitable material is used as the core of the electromagnetic coil arrangement.

magnetic field set up by a current flowing in a coil. Figure 10.7a shows a simple bell; when current flows the coil magnetises its core and the clapper arm is attracted to the coil core. The coil is called a *solenoid* in this application. When the arm moves across it opens the switch, cutting off the current supply, and if the material is of the correct type, demagnetisation occurs and the arm springs back to its rest position. When it does so the switch is closed, current flows again and the cycle is repeated. Each time the arm moves across, the clapper strikes the gong giving the familiar bell sound.

Figure 10.7b shows an electromagnetic relay which operates switches; when the coil current flows the arm moves in and the side arm outwards, opening or closing the switch as shown. In the

How the Ampere is Defined

As we have seen, the unit of electric current, the ampere, is the current flowing when one coulomb of electric charge moves per second. However, the ampere is not *defined* in this way. It is defined using the force that is set up when two conductors carrying current are placed a certain distance apart. We have seen above that when a conductor carrying current is placed in a magnetic field there is a force set up which is exerted on the conductor. If the magnetic field referred to is also set up by a conductor carrying current we have a situation in which a force exists between two conductors carrying current.

The ampere is that current which, when flowing in each of two infinitely long parallel conductors, situated in a vacuum and separated 1 metre between centres, produces a force between these conductors equal to 2×10^{-7} newtons per metre length.

Considering this formal definition further, 'infinitely long' can be taken to mean very long (there is a slight change in field patterns for short conductors), 'in vacuum' is necessary because the strength of the field or, to be precise, the flux density of the field, depends to some extent on the medium between the conductors (being slightly different in air, for example, than in a vacuum) and, finally, the rather strange figure 2×10^{-7} (meaning 2/10 000 000) is chosen deliberately so that relationships between other units depending on the ampere are as simple as possible.

ELECTROMAGNETIC INDUCTION

Many effects in science are reversible. Faraday, knowing that an electric current produced a magnetic field, wondered if the reverse were true. Did a magnetic field produce an electric current? He found that under certain circumstances it did. If a conductor is *moved* through a magnetic field or if a magnetic field is *moved* past a conductor (by moving a magnet past a conductor, for example) or if both field and conductor are stationary and the strength of the magnetic field is *changed*, then an e.m.f. is set up across the conductor, *provided that the magnetic field does not act in a direction parallel to the conductor*. If the conductor is part of a closed conductive circuit then current flows. The important necessary condition is that the magnetic flux *linking* the conductor should change. In the two cases involving movement, of either field or conductor, the magnetic field itself may be constant but, provided there is motion, the flux *linking* the conductor is changing from one instant to the next and the e.m.f. appears. The process is called *electromagnetic induction*.

If a coil is used to generate an e.m.f., then a number of conductors is involved and the *flux linkage* is defined as the product of flux and number of turns on the coil. Faraday discovered that *the value of the induced e.m.f. is proportional to the rate of change with time of the flux linkage*. If SI units are used this statement of proportionality becomes an equation as follows

induced e.m.f. = rate of change of flux linkage

= rate of change of (number of turns × flux)

and if the number of turns is constant and therefore does not change

induced e.m.f. = number of turns × rate of change of flux

This is a statement of Faraday's law of electromagnetic induction.

Another scientist, named Lenz, discovered that the voltage induced always acts in a direction so as to oppose what is causing it to be set up. This is illustrated in figure 10.8. Figure 10.8a shows a conductor moving into a magnetic field acting vertically downwards (the north pole or effective north pole of whatever is causing the field being at the top of the diagram). If current flows in the conductor due to the induced voltage across the conductor it will flow in a direction such that the conductor field set up by the current opposes the main magnetic field into which the conductor is moving. Now like fields repel, so the conductor field must act in the direction shown in figure 10.8b. Using the corkscrew rule mentioned earlier, current must flow into the paper, that is, away

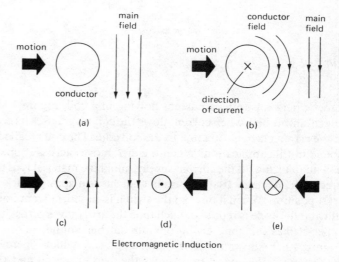

Electromagnetic Induction

Figure 10.8

from the observer, as indicated by the cross in the conductor. If current does *not* flow (because the conductor is not part of a complete circuit) the induced e.m.f. will act in a direction so as to cause the current direction given *if the circuit were complete*, that is, the end of the conductor nearest to us is positive. (Remember that the direction of current flow is that of conventional current, which flows from positive to negative.) The remainder of the figure shows current directions for other field directions and other directions of motion of the conductor. This method of determining the direction of action of an induced voltage is superior to most other methods.

Because of the opposing nature of the induced voltage it is called a *back e.m.f.* and is shown as negative when Faraday's law of electromagnetic induction is expressed mathematically, thus

$$E = -N \frac{d\phi}{dt}$$

where E is the induced e.m.f., N is the number of turns and $d\phi/dt$ is mathematical shorthand for 'the rate of change of flux with time'. (ϕ represents flux and t time, the symbol d being used in integral calculus, the mathematics of rates of change.) If the appropriate mathematics is not known by the student then the law should be remembered in words

$$\text{induced e.m.f.} = \text{number of turns} \times \frac{\text{rate of change of}}{\text{flux with time}}$$

(induced e.m.f. opposing the cause of the induction).

Rate of change may vary from instant to instant; dividing a flux change by the time taken for the change gives only the *average rate of change*. The same kind of comment applies to any rate of change—velocity, acceleration, etc. The average rate of change is only equal to the actual rate of change throughout the period of change if the rate of change is constant.

Example 10.3

The flux density of a magnetic field contained in a material of uniform cross-sectional area 0.09 m^2 changes from 0.1 T to 0.25 T in 0.08 s. If a 50-turn coil were situated in a magnetic field changing in this way, the field being at right-angles to the conductor, determine the average voltage induced across the coil.
Solution

$$\text{Flux density} = \frac{\text{flux}}{\text{area}}$$

$$\text{flux} = \text{flux density} \times \text{area}$$
$$= 0.1 \times 0.09 \text{ Wb}$$

when flux density is 0.1 T, and

$$\text{flux} = 0.25 \times 0.09 \text{ Wb}$$

when flux density is 0.25 T

$$\text{flux change} = (0.25 \times 0.09) - (0.1 \times 0.09)$$
$$= 0.0135 \text{ Wb}$$

which occurs in 0.08 s.

$$\text{Average rate of change of flux} = \frac{0.0135}{0.08}$$
$$= 0.169 \text{ Wb/s}$$
$$\text{Average e.m.f. induced} = 50 \times 0.169$$
$$= 8.44 \text{ V}$$

Example 10.4

A straight conductor 0.7 m long is moved with constant velocity at right-angles both to its length and to a uniform magnetic field. Given that the e.m.f. induced in the conductor is 5 V and the velocity is 20 m/s calculate the flux density of the magnetic field.
Solution   Suppose the required flux density is represented by B (the usual symbol), then

$$\text{flux} = B \times \text{area over which flux acts}$$

The conductor length 0.7 m moves 20 m every second, the situation being shown in figure 10.9. The area over which the flux causing the

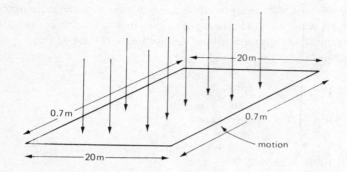

Figure 10.9

induction acts is thus 20×0.7, that is 14 m^2. Thus

flux $= 14B$

and this flux is cut every second, so that

rate of change of flux $= 14B \text{ Wb/s}$

(the velocity is constant). This is a single conductor, the number of turns therefore is one, so using

e.m.f. = number of turns × rate of change of flux with time
$5 = 14B$

and

$$B = \frac{5}{14}$$
$$= 0.357 \text{ T}$$

The flux density is 0.357 T.

Note that in these examples care was taken to say that the magnetic field linking the conductor was acting at right-angles to the conductor. If the field is *parallel* to the conductor then electromagnetic induction *does not* take place. A magnetic field may be considered to be a vector quantity made up of components (like velocity and force). If the field acts at any angle other than 90°, then it is the *component acting at right-angles* to the conductor which causes electromagnetic induction.

GENERATION OF AN ALTERNATING E.M.F.

Electromagnetic induction is used in the generation of voltage for domestic and industrial use. A simple generator is shown in figure 10.10. The similarity between this figure and figure 10.6, which shows a motor, is marked, the point being that generation is the motor principle applied in reverse. With a motor, the current and field are supplied and current results (voltage actually, but current if the circuit is closed). If the induced voltage across the coil is picked off using brushes and the *slip-rings* shown, the voltage between brushes will reverse polarity as the conductor moves first under the north pole then under the south pole. The magnitude also changes (as explained below) because the effective flux cutting the conductor at the right-angle required is changing as rotation takes place.

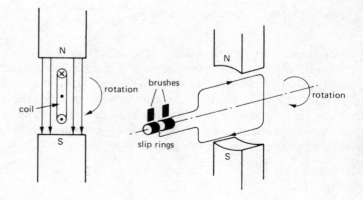

Figure 10.10

To obtain an understanding of why the induced voltage in the basic generator shown varies as it does, we must re-examine those factors which affect the size or magnitude of the voltage. For induction to take place, the magnetic field must cut the conductor at right-angles to its axis and must change in magnitude itself. The changing field may be obtained, as stated above, either by allowing the conductor to move through a field which is at right-angles (or has a component at right-angles) or the conductor may remain stationary and the field may change its value (due to whatever is causing it changing in some way). In the generator the conductor moves through a field, usually at constant speed. If the field were of constant value then the rate of change of field magnitude would be due to the speed of the conductor alone, which is constant, and the induced voltage would be constant. In the arrangement shown in figure 10.11 the field acts downwards in the diagram and the conductor rotates within the field.

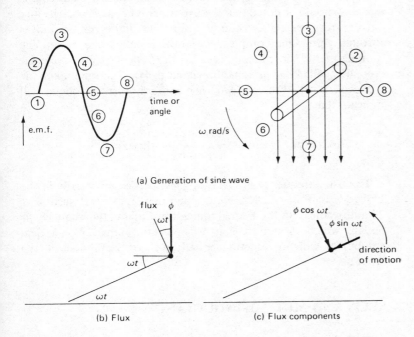

(a) Generation of sine wave

flux ϕ

(b) Flux

$\phi \cos \omega t$

$\phi \sin \omega t$

direction of motion

(c) Flux components

Figure 10.11

Consider any one of the conductors making up the coil. When the conductor is at position 1 (figure 10.11a) its direction of motion is parallel to the lines of force and no *cutting* of flux takes place. Thus no e.m.f. is induced and the voltage shown in the left-hand voltage–time graph is zero at position 1. At position 3 the conductor is moving at right-angles to the flux, cutting the full flux, and maximum voltage is induced as shown in the voltage–time graph at position 3. Between positions 1 and 3, at position 2 the conductor is cutting some but not all the flux and a voltage between zero and maximum is induced, as shown in the voltage–time graph. To further clarify this examine figure 10.11b, which shows the coil at an angle ωt to the horizontal and the flux ϕ acting vertically downwards in the diagram. This flux may be considered to have two components: one of magnitude $\phi \cos \omega t$ acting in the opposite direction to the direction of motion (see figure 10.11c) and $\phi \sin \omega t$ acting at right-angles to the direction of motion. It is the cutting of this latter component $\phi \sin \omega t$ which produces the induction of the e.m.f. The angle ωt is the angle made by the coil at any time t if the coil is rotating at an angular velocity of ω radians per second. Clearly, angle ωt varies from zero to 90° as the conductor moves from position 1 to position 3 in figure 10.11a and the voltage, which is given by

induced voltage = number of turns × rate of change of flux

will vary *sinusoidally*, that is, in accordance with the variation of the sine of angle ωt, since the flux varies in accordance with the sine of this angle. The sinusoidal shape of the voltage–time graph is shown in figure 10.11a and, as can be seen, the voltage rises to a maximum, falls to zero, then to a maximum value in the opposite direction and back to zero again. Positions 1 and 8 on the graph correspond to positions 1 and 8 of the conductor shown in the right-hand diagram.

An alternative way of looking at the situation is to leave the flux as it is, without components, and to consider the conductor velocity to have components, as shown in figure 10.12. Suppose the linear velocity of the conductor is v m, acting in the direction shown. Taking the coil angle with the horizontal to be ωt as before, where ω is the angular velocity of the coil, we can replace v with

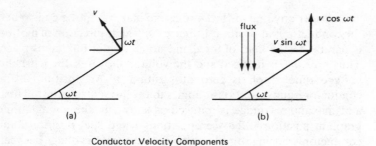

Conductor Velocity Components

Figure 10.12

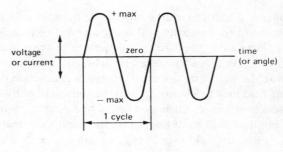

The Sine Wave

Figure 10.13

sin ωt, acting horizontally, and $v \cos \omega t$ acting vertically (see figure 10.12b). The velocity component which cuts the flux at right-angles as required is $v \sin \omega t$. We now have the situation of example 10.4, in which a straight conductor moves with constant velocity at right-angles both to its length and to a uniform magnetic field. If the conductor length is l metres then the area of flux swept through each second is $v \sin \omega t \times l$ square metres, that is, velocity of conductor × length; the rate of change of flux is flux density × area moved through per second so that

rate of change of flux = flux density × $v \sin \omega t \times l$

and as before we see that the rate of change of flux is varying sinusoidally so that the induced voltage will do likewise.

Sinusoidal variation of voltage and current will be considered in more detail later.

The basic generator shown is used for relatively small values of voltage; in large machines where voltages of the order of 11 kV are generated, difficulty is experienced in picking off the voltage at the slip-rings. Therefore in these machines the conductors are stationary and the magnetic field is set up by a rotating coil. The field coil is at lower voltage so the same problem is not encountered.

SINUSOIDAL VARIATION

A sinusoidally varying voltage–time or current–time graph is shown in figure 10.13. As described above the voltage or current (produced by a sinusoidally varying voltage) rises from zero to a positive maximum then falls to zero, to rise to a maximum value in the opposite direction and return to zero before the process begins again. It should be noted that voltage actually changes polarity and current reverses its direction of flow as the variation moves through zero. One complete variation is called *one cycle*. The number of cycles per second is called the *frequency*. One cycle per second is called *one hertz*, abbreviated to Hz. The time taken for one cycle is called the *periodic time* and a moment's thought will indicate that

$$\text{periodic time (seconds)} = \frac{1}{\text{frequency}} \text{(hertz)}$$

The usual frequency of voltages generated commercially in the United Kingdom is 50 Hz, giving a periodic time of 1/50 second or 20 milliseconds. In the United States and Canada the frequency is 60 Hz, which is one reason why certain items of electrical equipment will not function equally well on both sides of the Atlantic.

SELF- AND MUTUAL INDUCTANCE

A voltage is induced across a conductor whenever it is situated in a magnetic field which is changing. In the generator, conductors are deliberately placed in a changing field in order to generate voltage; but the effect of voltage induction is not confined only to generators. Since an electric current sets up its own magnetic field surrounding the conductor in which the current flows, whenever a current changes there will be an induced voltage. Further, the induced voltage will act so as to oppose whatever is causing its induction. The effect is called self-inductance when the induced voltage is set up due to a conductor's own field.

All conductors have self-inductance. How large the effect is depends on a number of variables: the rate of change of current and thus of flux causing the induction, the strength of the flux caused by the current (as determined by whether or not the conductor is wound in coil-form) and also by the medium surrounding the conductor. If we wish the self-inductance to be large, as we do on occasions, the conductor should be wound in the form of a coil and the coil placed on a ferromagnetic core. This increases the flux produced by the current so that when the current changes the flux will be changing from a larger value and thus the rate of change of flux will also be larger.

When considering self-inductance, since induced voltage depends on rate of change of magnetic flux and magnetic flux is set up by electric current then

induced voltage depends upon rate of change of current

or more precisely

induced voltage is directly proportional to rate of change of current

and writing this as an equation

induced voltage = a constant × rate of change of current

The constant is called the coefficient of self-inductance of the conductor and has the symbol L. It is measured in *henrys* (note the plural, the singular being a henry, unit symbol H), where one henry is the coefficient of self-inductance of a conductor when a current changing at the rate of one ampere per second causes a voltage of one volt to be induced across the conductor ends. The value of the coefficient of self-inductance depends on the factors discussed above, namely, how the conductor is wound, on what it is wound (the core material) and one other factor: the state of magnetisation of the core. If a ferromagnetic material is used and it is near saturation, so that flux cannot increase, this will affect the value of flux and thus its rate of change.

The effect of self-inductance is noticeable particularly when circuits associated with magnetic fields are opened and closed. Equipment such as motors and relays require the setting up of a magnetic field for them to function correctly. Accordingly, whenever the circuit of the motor (for the field if this is being set up electromagnetically) or relay is opened, a voltage is induced which tries to maintain the state of affairs prior to the opening, that is, the back e.m.f. will try to keep current flowing. This effect is observable as sparks appearing across switch contacts.

Example 10.5

The self-inductance of a coil is 0.5 H. Find the back e.m.f. induced when a current of 2 A is reversed in the coil in 20 ms. Assume a constant rate of change of current with time.
Solution

$$\text{Back e.m.f.} = \text{coefficient of self-inductance} \times \frac{\text{rate of change}}{\text{of current}}$$

Since the current changes from $+2$ A to -2 A, the effective change is 4 A in 20 ms. The average rate of change is therefore $4 \div (20 \times 10^{-3})$ which equals 200 A/s, and

$$\begin{aligned}\text{back e.m.f.} &= 0.5 \times 200 \\ &= 100 \text{ V}\end{aligned}$$

The value here is fairly high for an induced voltage; much higher values can be obtained, of the order of kilovolts, for use in, for example, internal-combustion engine spark plugs.

If two conductors are placed so that the magnetic field set up by the current in one links the other conductor, then, when the current is changed, an induced voltage will be set up by the current across both conductors. The effect of a changing current setting up an induced voltage across an adjacent conductor is called *mutual inductance*. The induced voltage across the second conductor is related to the changing current in the first conductor by the equation

$$\text{induced voltage across second conductor} = M \times \text{rate of change of current in first conductor}$$

where M is the coefficient of mutual inductance between the conductors. If a current changing at the rate of 1 A/s in the first conductor induces a voltage of 1 V across the second, then the coefficient of mutual inductance between the conductors is 1 henry, the same unit as for the coefficient of self-inductance.

Example 10.6

The mutual inductance between two coils is 0.3 H. Find the e.m.f. induced across one coil when the current in the other changes uniformly from 6 A to 3 A in 10 ms.

Solution

Change in current = 3 A in 10 ms

$$\text{Rate of change of current} = \frac{3}{10 \times 10^{-3}}$$

$$= 300 \text{ A/s}$$

e.m.f. induced = coefficient of mutual inductance
× rate of change of current

$$= 0.3 \times 300$$

$$= 90 \text{ V}$$

The coefficient of mutual inductance between coils depends on the medium in which the coils lie, which in turn determines the flux for a particular value of current. The mutual-induction effect is used in transformers where it is possible, using closely placed coils wrapped on a magnetic medium, to change voltage levels. The voltage to be changed, called the *primary* voltage, is connected to one coil producing a current and thus an interconnecting flux. The changing flux around the second coil induces a voltage called the *secondary* voltage. It can be shown that the ratio of primary voltage to secondary voltage is equal to the ratio of the number of turns on the primary coil to the number of turns on the secondary coil.

A Way of Defining the Weber

Since induced voltage across a coil is connected to the rate of change of flux around the coil by the equation

$$\text{induced e.m.f.} = \text{number of turns} \times \text{rate of change of flux}$$

if the voltage across a one-turn coil is 1 V and this voltage is induced by a changing flux, then the flux must be changing at the rate of 1 weber per second. One weber is thus one volt-second.

ASSESSMENT EXERCISES

Long Answer

10.1 Draw diagrams to show the magnetic-field pattern of (a) a bar magnet, (b) a straight conductor carrying current. How may the field of a current-carrying conductor be increased? Describe briefly two devices in everyday use which use electromagnetism as a means of operation.

10.2 State Faraday's law of electromagnetic induction. Describe, using diagrams, how a back e.m.f. is set up whenever an electric current flowing in a conductor is modified. Directions of currents and voltages must be shown in the diagrams.

10.3 (a) Briefly explain what is meant by 'electromagnetic induction'. (b) A straight conductor 0.75 m long is moved with constant velocity at right-angles both to its length and to a uniform magnetic field of density 0.6 T. The conductor velocity is 10 m/s. Calculate the e.m.f. induced in the conductor.

10.4 (a) Explain the term 'back e.m.f.'. (b) A current of 1.5 A is reversed at a uniform rate in 25 ms. The coil in which the current flows has a coefficient of self-inductance of 2.4 H. Determine the back e.m.f. induced across the coil.

10.5 The current through a coil changes uniformly from 0 to 0.75 A in 40 ms resulting in a back e.m.f. of 1000 V. Calculate the coil self-inductance.

10.6 (a) Explain what is meant by mutual inductance. (b) The current in a coil changes uniformly from 2 A to 8 A in 0.7 s. The e.m.f. induced across an adjacent coil is found to be 50 V. Determine the mutual inductance between the coils.

10.7 The mutual inductance between two coils is 0.8 H. Find the e.m.f. induced across one coil when the current through the other changes uniformly from 0.1 A to 1.5 A in 1.4 s.

10.8 (a) State the relationship between back e.m.f., coefficient of self-inductance of a coil and the rate of change of current flowing in the coil. (b) A current in a coil changes from -2 A to $+1$ A in 0.2 s and the back e.m.f. is 157 V. Calculate the coefficient of self-inductance of the coil.

10.9 (a) Explain briefly the principle of the simple generator. (b) A coil of 150 turns cuts a magnetic field of density 0.25 T at right-angles at a speed of 10 m/s. The induced voltage is 250 V. Determine the effective length of the conductors making up the coil.

10.10 A current of 1 A flowing in a coil of 750 turns sets up a flux of 0.11 mWb. The current is reversed uniformly in 2 ms. Calculate the induced e.m.f. and the coefficient of self-inductance of the coil.

10.11 A conductor of length 250 mm is moved at a constant velocity of 25 m/s at right-angles to its length and to a magnetic field of density 0.8 T. The conductor is connected externally to a circuit through which a current of 20 A flows. Calculate (a) the e.m.f. induced across the conductor, (b) the retarding force set up by the current flowing in the conductor, (c) the power necessary to maintain motion.

Short Answer

10.12 Calculate the force per metre length acting on a conductor carrying 2.2 A situated at right-angles to a magnetic field of flux density 0.15 T.

10.13 The force per metre length on a conductor carrying 0.4 A placed at right-angles to a magnetic field is 10 N. Determine the flux density.

10.14 Determine the average value of the e.m.f. induced across a 75-turn coil situated at right-angles to a magnetic field which changes at the rate of 0.03 Wb/s.

10.15 State Faraday's law of electromagnetic induction.

10.16 Calculate the periodic time of a voltage varying sinusoidally at 50 cycles per second

10.17 Calculate the frequency of a sinusoidally varying current which has a periodic time of 15 ms.

10.18 Define the coefficient of self-inductance.

10.19 Define the coefficient of mutual inductance.

10.20 The back e.m.f. induced across a coil when a 1.5-A current through it is reversed in 0.6 s is 50 V. Determine the coil self-inductance (assuming constant rate of change of current).

10.21 The mutual inductance between two coils is 0.7 H. Find the e.m.f across one coil when the current in the other changes at 1.5 A/s.

Multiple Choice

10.22 The unit of magnetomotive force is the
A. tesla   B. weber   C. ampere-turn   D. henry

10.23 The unit of magnetic flux is the
A. tesla   B. weber   C. ampere-turn   D. henry

10.24 The unit of magnetic flux density is the
A. weber   B. ampere-turn   C. tesla   D. henry

10.25 The unit of the coefficient of self-inductance is the
A. henry   B. weber   C. tesla   D. ampere-turn

10.26 The current flowing in a wire placed in a magnetic field of flux density 0.2 T when the force per metre length of the conductor is 0.6 N is equal to
A. 3 A   B. 0.12 A   C. 0.33 A   D. incalculable without further information

10.27 The voltage induced per turn of a coil subjected to a changing magnetic field is equal to
A. flux/time   B. current/time   C. rate of change of flux with time   D. rate of change of flux with current

10.28 The voltage induced across a coil in terms of its coefficient of self-inductance L is equal to
A. $L \times$ flux/time   B. $L \times$ current/time   C. $L \times$ rate of change of current with time   D. $L \times$ rate of change of flux with time

10.29 The back e.m.f. induced across a coil of self-inductance 0.4 H when a 1.5-A current is reversed in 0.2 s has an average value of
A. 3 V   B. 6 V   C. 53.3 mV   D. 26.67 mV

10.30 The e.m.f. across a coil is 4 V when the current through an adjacent coil is changing at the rate of 2 A/s. The mutual inductance between the coils is
A. 2 H   B. 8 H   C. 0.5 H   D. 1 H

10.31 The periodic time of a 60-Hz sine wave is (in seconds)
A. 60   B. 0.0167   C. 16.67   D. 0.00833

11 Chemical Reactions

OBJECTIVES

All the objectives should be understood to be prefixed by the words 'The expected learning outcome is that the student . . .'

A3 Describes the atomic structure of matter and the nature and formation of crystals.

3.1 Describes the atom as the basic building block of matter.

3.2 Describes the molecule as an independent group of atoms bonded together.

3.3 Explains the terms elements and compounds in terms of atomic composition and distinguishes compounds from mixtures.

3.4 Gives three examples of each of the following (a) elements (b) compounds (c) mixtures.

3.5 Defines a solution.

3.6 Defines a suspension.

3.7 Defines solubility.

3.8 Lists common factors influencing solubility of a solid in a liquid.

3.9 Defines a saturated solution.

3.10 Recognises that crystals have flat faces and specific angles.

3.11 Describes the process of crystallisation from a solution.

3.12 Describes metals as polycrystalline substances.

3.13 Recognises that alloys can be solid solutions.

G18 Describes oxidation.

18.1 States that air is a mixture mainly of oxygen and nitrogen.

18.2 Describes how a substance such as copper gains mass when heated in air and that oxygen is taken from the air by the copper.

18.3 Describes how substances burning in air combine with oxygen.

18.4 Describes an oxide as a compound of an element and oxygen.

18.5 Describes how oxygen and water are involved in rusting.

18.6 Discusses the damage done by rusting.

G19 Describes the effects of electricity on substances.

19.1 Describes metals (and carbon) as good conductors, in the solid state, of electricity.

19.2 Describes electroplating.

19.3 Explains conduction in liquids as due to ions, charged parts of a molecule.

19.4 Describes a simple cell.

19.5 Uses the electrochemical series to make qualitative predictions about pairs of metals in cells.

19.6 Relates the series to the pattern of the reactivities of metals with oxygen.

19.7 Describes corrosion as 'simple cell' action.

19.8 Discusses the damage done by corrosion.

G20 Describes and gives a simple explanation of chemical reactions.

20.1 Describes acidity/alkalinity by means of indicators.

20.2 Describes a base as a substance which removes the acidic properties of acids.

20.3 Describes an alkali as a soluble base which yields hydroxyl ions.

20.4 Describes chemical reactions as interactions between substances which involve a rearrangement of atoms.

20.5 Prepares a salt by neutralisation of an acid.

20.6 Describes an acid as a compound which yields hydrogen ions.

20.7 Describes a chemical equation as representing a re-arrangement of elements.

The word 'science' comes from a Latin word meaning 'to know'. Science is knowledge, particularly knowledge obtained from observation, study and experimentation. Scientists spend their time attempting to answer questions—why things are as they are, why materials behave as they do, why events occur, and so on. To answer such questions scientists suggest theories and, hopefully, go on to prove or disprove them by experiment. They attempt to explain observed facts and perhaps to predict happenings not yet observed. Scientific knowledge is increasing rapidly as we become more able to test theories and draw definite conclusions. In many areas, one of which we are about to examine, the theories we have do not explain everything. They are a great improvement perhaps on the ideas of earlier days when, for example, all matter was believed to be made up of varying proportions of fire, air, earth and water or when heat was believed to be a substance called 'caloric', but they are nevertheless ideas which are subject to change or even rejection. It is wise to bear this in mind with all scientific theory and accept it for what it is—the best explanation *so far* of scientific fact.

CHANGES

Under normal conditions of temperature and pressure—'normal' being what we experience in everyday life—oxygen and hydrogen are gases. However, combined together in the correct proportions, they form water. This is an example of a *chemical change*. When a piece of soft iron is magnetised its behaviour in relation to other pieces of iron is changed, but the iron does not change. This is an example of *physical change*. The main difference between chemical and physical changes is that a chemical change results in a new substance being formed whereas a physical change does not.

All substances are made up of very tiny parts called *atoms* arranged in groups called *molecules*. Later we shall define these terms more precisely. For the moment we can say that in a chemical change the atoms separate and change themselves in new ways to form molecules of a different substance. In a physical change the molecules of a substance remain intact and no new substance is formed.

ELEMENTS, COMPOUNDS AND MIXTURES

The number of substances in the world, naturally occurring or man-made, appears limitless. All these substances, however, contain one or more of a number of basic materials called *elements*. An element is a substance which cannot be split up into anything simpler by a chemical change. Water, for example, can be split up into hydrogen and oxygen; hydrogen and oxygen cannot be further split into different substances and so are elements. We know of just over one hundred elements, ranging from common ones known for many years—hydrogen, helium, oxygen, sodium—to more recently discovered ones such as berkelium, einsteinium, nobelium and lawrencium.

Elements themselves are made up of very tiny parts called atoms. An atom is the smallest possible part of an element which can exist, retain the chemical individuality of the element and take part in a chemical change. By 'chemical individuality' we mean the way in which the element behaves chemically. Taking the element carbon as an example, we can take a piece of the element and divide it continuously until we reach the carbon atom. Although we could then break up the atom, if we did we would no longer have carbon.

When elements combine chemically their atoms interlink to form molecules of a new substance called a *compound*. A compound is a substance which contains two or more elements combined so that their properties are changed. Elements or compounds may be mixed together without chemical change so that no interlinking of atoms takes place. The result is called, simply, a *mixture*. As an example, iron filings may be mixed with sulphur particles and the result is a mixture of black and yellow particles. A magnet attracts the iron filings which are still chemically separate from the sulphur. If, however, the mixture is heated the result is a compound called ferrous sulphide which is black, has no separate iron and sulphur particles and is unaffected by a magnet. A chemical change has taken place for the iron–sulphur mixture to become the ferrous sulphide compound. Some common examples of compounds and mixtures are given in table 11.1.

Table 11.1

Compounds	Mixtures
Salt (sodium and chlorine)	Salt and water
Water (hydrogen and oxygen)	Sand and water
Sand (silicon and oxygen)	Alcohol and water
Caustic soda (sodium, hydrogen and oxygen)	Sand and salt

ATOMS AND MOLECULES

An atom is the smallest part of an element that can exist and retain the chemical individuality of the element. It is the smallest part of an element that can take part in a chemical change. A molecule is a combination of atoms, either of the same element or different elements. Some elements when existing freely on their own exist as multi-atom molecules rather than single atoms. Hydrogen and oxygen, for example, have molecules each containing two of the appropriate atoms. Metallic elements and inert gases have atoms existing on their own. A molecule is the smallest part of a substance which can have a *separate stable existence*. Oxygen on its own consists of two-atom molecules (called diatomic) but in a chemical change with hydrogen to form water one *atom* of oxygen will combine with two atoms of hydrogen to form one *molecule* of water. Here we have the smallest part of oxygen which can take part in a chemical change, that is, one atom of oxygen, combining with hydrogen to form the smallest part of water which can have a separate stable existence, namely a molecule of water.

If water is split up into its component elements, hydrogen and oxygen, the atoms of each gas combine in pairs to form diatomic molecules of each gas. A hydrogen or oxygen atom will not exist on its own but will combine with a similiar atom to form a molecule of the element, the molecule again being the smallest part of hydrogen or oxygen which can have a separate stable existence.

Basic Atomic Structure

The ideas currently accepted of the structure of an atom originated with John Dalton, a Manchester scientist of the early nineteenth century, and were further developed by Lord Rutherford and Niels Bohr. As was stated earlier, these ideas are subject to continual change and revision as scientific research and experiment continues. Basic atomic theory remains more or less the same but even this could be revised in the light of any new and startling discoveries.

The simplest picture of the atom describes it as consisting of a central nucleus around which *electrons* orbit in much the same way as the planets orbit the Sun in our solar system. The electron at its simplest can be regarded as a very tiny particle with an electric charge which we call negative. (In fact the electron has been observed to behave differently from a particle on some occasions but this need not concern us at the moment.) The atomic nucleus is large and relatively heavy compared to the electron but the word 'relatively' is vitally important, since the size and weight of an electron are very very small indeed. The nucleus contains two further kinds of particle: protons and neutrons. The proton may be regarded as a small, light particle with an electric charge equal and opposite to that on the electron. This charge is called positive. The neutron is a relatively heavy particle and has no charge. Over all the atom is electrically neutral, that is, it contains the same number of protons as electrons; this number is called the *atomic number* of the element of which the atom is part.

The electrons orbiting the nucleus are distributed in a number of fixed orbits or *shells*, depending on the complexity of the atom. The simplest atom is that of hydrogen and has one electron orbiting the nucleus; the nucleus contains one proton. Helium has two electrons per atom and each atomic nucleus has two protons. The two electrons occupy the same shell and orbit at the same distance, as shown in figure 11.1. As far as we are aware, the first shell of atoms in general can have up to two electrons and no more, because the next more complex atom is lithium, with three electrons per atom, two in the first shell and one in the next shell, as shown in figure 11.1. The second shell can take up to eight electrons, the materials so formed being in order lithium, beryl-

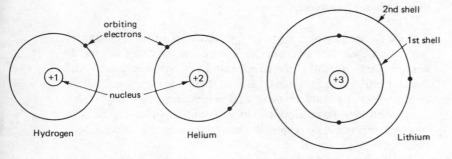

Figure 11.1

lium, boron, carbon, nitrogen, oxygen, fluorine and neon. The next atom in order of complexity is sodium, which has eleven electrons, two in the shell closest to the nucleus, eight in the next shell, and one in the third shell.

In general it has been found that if we denote the shell number by n, the maximum number of electrons in the shell can be $2n^2$, that is, shell 5 can have 2×5^2 or 50 electrons in it. At the time of writing the element with 50 electrons in its fifth shell has not yet been discovered; there may of course not be one.

As stated, atoms are neutral but electrons can be removed or added to certain atoms without changing the basic element and such atoms should then strictly be called *ions*.

SOLUTIONS

If a small amount of sugar or salt is put in water and the liquid stirred or shaken, the solid disappears leaving a clear liquid called a *solution*. Small amounts of certain other materials, such as chalk, do not behave in this way (that is, do not 'disappear') and the resulting mixture is called a *suspension*. If a material dissolves in the manner of sugar or salt in water, it is called *soluble*. If a material does not dissolve and a suspension is obtained, the material is called *insoluble*. Soluble material when dissolved in a liquid is called the *solute*, the liquid being called the *solvent*. Water

is not, of course, the only possible solvent. Petrol, for example, dissolves fat, in which case fat is the solute and petrol the solvent. Solutes may also be materials other than solids. They may be liquids or gases as, for example, when liquid acid is diluted by adding water or when the gas hydrogen chloride is dissolved in water to form hydrochloric acid. Similarly solvents may be materials other than in liquid form. Certain metallic alloys composed of two metals may be considered to be solid solutions. Alloys are described in more detail later.

Summarising, then, a solution is a perfect mixture of two or more substances, whereas a suspension is a mixture of a liquid and a finely divided but insoluble solid. Note that in both cases we obtain a mixture, and that usually the change is physical rather than chemical, that is, no new substance is formed. Generally, a solute may be reclaimed from a solution without undue difficulty, usually by evaporation of the solvent if it is water. It cannot be reclaimed by filtering, however, which is a method used for reclaiming the insoluble material from a liquid suspension.

When a solution is made it is found that there is a limit to the amount of solute that can be added to the solvent. After this point is reached the solution is *saturated* and any excess solute is not dissolved. The point at which a solution becomes saturated depends particularly on temperature. The *solubility* of a soluble material in a particular solvent at a given temperature is the maximum mass of the material that will dissolve in a given mass of the solvent, in the presence of excess of the solute at the given temperature. Solubility depends on temperature, on the solute and the solvent. A graph plotting solubility against temperature is called a solubility curve. It should be noted that solubility in general does not necessarily rise with temperature but may in fact fall as the temperature of the solvent is increased. Solubility of solids tends to rise and that of gases tends to fall as temperature is increased. Pressure of the surrounding atmosphere or other gas hardly effects solid solubility but does increase that of gases.

Certain metallic elements can be mixed to produce what is effectively a solution called an *alloy*. The two main groups of alloys are ferrous (containing iron) and non-ferrous. The most familiar ferrous alloy is steel, which contains iron and a very small proportion of carbon (about 0.2 per cent). Stainless steel contains in addition an amount of chromium which helps prevent rusting. (Rusting is considered later.) Other metals used in alloying steel include manganese, tungsten, cobalt, molybdenum and aluminium. Of the non-ferrous alloys one of the best known is probably duralumin, made up of aluminium, copper and magnesium. Others include a number of magnesium, copper, tin, zinc and lead alloys used in a variety of ways in industry. The most widely used non-ferrous alloys in the aircraft industry are those based on aluminium, which is a very light material but by suitable addition of other materials it can produce a tough, long-lasting, corrosion-free alloy.

CRYSTALS

A crystal is a solidified form of a substance made up of plane (flat) faces in a symmetrical arrangement. Some typical examples are shown in figure 11.2. There are many examples of crystalline form which occur naturally; some of the more common include diamond, common salt, graphite, quartz and fluorite. The definite geometrical shape of a crystal is determined particularly by the arrangement of the molecules or atoms of which it is composed.

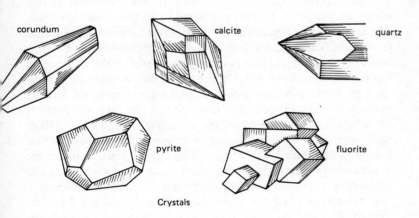

Crystals

Figure 11.2

P.S.—K

Different arrangements of the same atoms can produce different properties of what is basically the same material, a good example being graphite and diamond, both of which are carbon but the former is used as a lubricant and the latter as an abrasive.

The properties of perfect crystals are that they are bounded by plane faces, edges and corners, and have specific angles formed between faces and edges, that is, they have a definite geometrical shape and they can be easily divided into smaller crystals. All crystals of a particular material have the same form. Substances which are non-crystalline like glass, for example, are called *amorphous*. It is of course possible to cut glass into a crystal form with flat edges and so on but this is not then due to a naturally occurring arrangement of the glass molecules.

Crystals may be formed by allowing a hot saturated solution to cool, by allowing certain liquids to solidify and when gases solidify without passing through the liquid state. In the first method it is usual to start the process by suspending a tiny crystal of solute in the saturated solution; on cooling, the solute comes out of solution and is deposited on the tiny crystal already present. The crystal then grows, each layer being added in a definite geometrical form the same as the layers already present. The solute comes out of solution as the solution is cooled because solubility of a solid falls as temperature is decreased. Examples of the other methods described above are ice crystals formed when water freezes and hoar frost on a window when water vapour turns directly to ice.

Most materials, including metals, are *polycrystalline*. This means that they consist of a large number of crystals joined at the boundaries, the greater the number of boundaries the stronger being the material. A metal may be hardened by cooling quickly which produces a large number of crystals and thus many boundaries. The process of *annealing* in which a metal is cooled slowly causes larger crystals, less boundaries and thus a softer material as an end product.

OXIDATION

The air we breathe is a mixture of several gases including nitrogen, oxygen, carbon dioxide, argon, water vapour and, in urban environments, sulphur dioxide. The two main gases, nitrogen and oxygen, occupy between them about 99 per cent of any given volume of air, the remaining constituents being present only in very small proportions. Oxygen occupies about 21 per cent by volume and 23 per cent by weight of the atmosphere and is the commonest of all elements in the Earth's crust as a whole, making up almost half of it in one form or another.

Oxygen was discovered as a separate element, but not named as such, by two scientists working independently in 1774, Joseph Priestley in England and C. W. Scheele in Sweden, who were investigating what happens during heating or burning of materials. The modern theory of burning is attributed to Antoine Lavoisier, a Frenchman, who, after meeting Priestley in Paris in 1774, confirmed that oxygen is the part of air which combines with burning materials to cause an increase in weight. It was Lavoisier who gave it the name oxygen.

Before 1774, scientists had been at odds over what actually happens when something is burnt. Some materials, such as paper, wood and coal, appear to lose weight when they burn and others, such as copper, magnesium and iron filings, gain weight after heating. One explanation was an imaginary material called 'phlogiston' which was supposed to be present in all matter that could be burnt to a greater or lesser extent. On burning, the phlogiston escaped and the substance lost weight. The idea of phlogiston did not, however, explain an increase in weight and to get round this it was even assumed that sometimes phlogiston could have negative weight! We now know that in fact all materials when they are heated or burned in air absorb oxygen and gain in weight. The apparent loss of weight when wood, coal or other materials which burn to an ash are burnt, is because of the smoke given off and lost. Controlled heating where no smoke is lost has shown conclusively that these materials too absorb oxygen from the air. All materials require the presence of oxygen to a greater or lesser degree for burning to take place. (The observation that gunpowder will burn apparently without oxygen is explained by the fact that there is a certain amount of oxygen present in saltpetre, which is one of the constituents of gunpowder.) One of the tests of oxygen is its ability to relight a glowing splint, so if relighting occurs on insertion of a splint into a gas we know oxygen is present.

Oxygen is a most important element. It is colourless, odourless and tasteless, slightly soluble in water (which is essential for water life), can be liquefied at $-183\,°C$ and has approximately the same density as air. Industrial use is in chemical processing and, together with acetylene, in oxy-acetylene cutting and welding of metals. Medical use is made of oxygen to assist breathing in high altitude flying and climbing, submarines and mines.

When a material is heated in air it absorbs oxygen and a chemical reaction takes place resulting in the formation of the *oxide* of the material. Some examples are copper to form copper oxide, magnesium to form magnesium oxide, carbon to form carbon dioxide (in this case each molecule of carbon dioxide contains two atoms of oxygen, hence the prefix 'di' in dioxide), phosphorus to form phosphoric oxide, and so on.

Oxide formation, especially on metals, may or may not be advantageous. In some cases the oxide once formed protects the metal underneath, but in others, where the metal is being used as a conductor of electricity for example, and the oxide of the metal is a non conductor, bad electrical connections may ensue. Another everyday example of oxide formation being distinctly disadvantageous is the process of *rusting*. We are all familiar with the changes that take place on the surface of iron or iron-based metals which are left out of doors unprotected by paint or other material. The reddish-brown flakes of rust which form are in fact hydrated iron oxide. Rusting is a similiar process to burning in that the iron absorbs oxygen from the atmosphere. However, in this case heat is not required but water must be present as well as traces of carbon dioxide. Rusting is quite a complicated process, in which ferrous carbonate is first formed, then oxidised to ferric carbonate, which absorbs water to give hydrated ferric oxide (rust) and some carbon dioxide. Rust is a major problem in industry and also in domestic surroundings since much of our construction work in machines and buildings uses iron or iron-based materials. Left to itself the rusting process continues until all or most of the iron in the vicinity is consumed and holes or major areas of weakened metal are left behind. Any suitable material to prevent the metal remaining in close contact with air and water may be used as a rust preventative.

TABLE 11.2

Anode Material	Cathode Material	Electrolyte	Reaction at Anode	Reaction at Cathode
Platinum or carbon	Platinum, copper or carbon	Dilute sulphuric acid	Oxygen released Solution becomes less dilute as water is decomposed	Hydrogen released
Carbon	Carbon	Hydrochloric acid	Chlorine released	Hydrogen released
Copper	Copper, platinum or carbon	Copper sulphate	Anode dissolves Used for copper refining (cathode is pure copper, anode is impure)	Copper deposited on cathode
Carbon	Carbon or platinum	Sodium chloride (salt solution)	Chlorine released	Hydrogen released

Commonly used materials include paints, oils and resins.

ELECTROLYSIS

Materials may be classified according to whether or not they conduct electricity, that is, whether or not it is possible to pass an electric current through them. Thus we have good conductors and bad conductors, or insulators, and a group in between which is neither good nor bad. The word conductor in this context normally brings to mind metals such as copper, aluminium, silver, gold and alloys of these and other materials, but it is equally possible to pass an electric current through certain liquids by a process called *electrolysis*.

As was stated earlier an atom or group of atoms is normally electrically neutral; they contain equal numbers of positive and negative charges. If additional electrons are added to or taken away from these atoms or groups of atoms, in such a way as not to change the nature of the original material, the atoms or groups then become positive or negative *ions*. When certain soluble substances are dissolved in water, ionisation occurs spontaneously and positive and negative ions wander aimlessly through the liquid, the degree of 'wandering' or 'drifting' being determined by temperature among other things. (The word ion is derived from a Greek word meaning 'wanderer'.) Such solutions are called *electrolytes*. An example is sulphuric acid which, on dilution, produces positive hydrogen ions and negative sulphate ions. (The chemical name for sulphuric acid is hydrogen sulphate; it contains hydrogen, sulphur and oxygen.)

Now if electrically conducting rods, called *electrodes*, are placed in the electrolyte and one electrode is made electrically positive (the *anode*) and the other is made electrically negative (the *cathode*) by connection to a battery, the ions move in a definite direction — positive ions to the cathode, negative ions to the anode — and a direct (unidirectional) electric current flows.

Chemical changes take place within the electrolyte and at the electrodes. It is found that hydrogen and/or metals are liberated at the cathode and non-metals are liberated or metals are dissolved at the anode. Some examples are given in table 11.2.

Electrolysis is widely used in *electroplating*, which is the deposition of one material on another. In silver plating, for example, the electrolyte is silver nitrate and the anode is silver; the cathode is the article to be silver plated. During the electrolysis process the anode dissolves and the cathode becomes coated in silver. Another widely used plating process uses chromium to provide an attractive relatively tarnish-free surface on cars and other machines.

THE SIMPLE CELL

In the process of electrolysis, electrical energy is converted into chemical energy. As is often the case in nature, the process is reversible and chemical energy may be converted to electrical energy. The conversion is used in cells and batteries (discussed in more detail in a separate chapter). In this chapter we shall examine the action of the simple cell and its implications.

One form of simple cell consists of a zinc electrode and a copper electrode immersed in a solution of dilute sulphuric acid, as shown in figure 11.3.

In the simple cell the zinc electrode dissolves, releasing zinc ions into the solution and leaving the electrode negative. Positive hydrogen ions travel to the copper electrode making it positive. Thus an electromotive force is set up between the copper electrode (the anode) and the zinc electrode (the cathode). The e.m.f. is about 1 volt. If the electrodes are connected to a conductor, electrons travel from zinc to copper through the conductor and a direct electric current is established. When this happens more zinc is dissolved, more hydrogen is liberated from the acid and the p.d. is

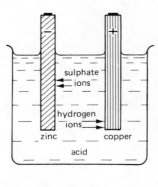

Simple Cell

Figure 11.3

maintained across the conductor. The action stops when one or more of the following occur

(1) the zinc dissolves completely

(2) the acid becomes inoperative when hydrogen ions are released to travel to the copper, sulphate ions travel to the zinc and the acid becomes less and less effective

(3) an excess of hydrogen forms at the anode insulating it from the electrolyte.

The third possibility is called *polarisation* and is the reason why the simple cell is not used practically. For practical use a *depolarising agent* is employed (see chapter 12).

THE ELECTROCHEMICAL SERIES

In the examples of electrolysis given earlier, we saw that with a copper sulphate electrolyte the metal copper is deposited on the cathode, whereas with a sodium chloride electrolyte the metal sodium is not deposited at the cathode but instead hydrogen is released. When metals are examined as a whole it is found that about as many are deposited during electrolysis as are not, hydrogen always being released instead of those which are not. If a simple cell is made up of each metal in turn, using an appropriate electrolyte containing ions of the metal and a second electrode of hydrogen (achieved by bubbling the gas against a suitable metal conductor, usually platinum) it is found that those metals which are deposited during electrolysis are *negative* with respect to the hydrogen electrode, whereas those metals which are *not* deposited during electrolysis become *positive* with respect to the hydrogen electrode. Further, the e.m.f. of each simple cell differs according to the nature of the metal electrode.

If metals are arranged in order according to the size and polarity of the e.m.f. produced when they are used in a simple cell arrangement with hydrogen, the result is the *electrochemical series*. The series proves a most useful guide as to how metals react not only with each other but with non-metals.

The series of the more common metals is given below.

Potassium Sodium	Alkali metals
Calcium Magnesium	Alkaline earths
Aluminium Zinc Iron Tin Lead Copper Mercury	Heavy metals
Silver Gold Platinum	Noble metals

As shown, they are grouped into *alkali*, *alkaline earths*, *heavy* and *noble* metals. Each group has similar properties to some extent. The noble metals are referred to as being at the 'low' end of the series whereas the alkali metals are at the 'high' end.

Closer examination of the behaviour of metals shows that the electrochemical series can be used as a guide to what to expect under various conditions.

(1) Metals tend to displace those lower than themselves in the series from a solution containing ions of the lower metal. For example, if an iron nail is dropped in copper sulphate solution it becomes covered in copper. Iron is higher in the series so it displaces the copper from the copper sulphate solution; the displaced copper then covers the nail.

(2) The higher the metal is in the series the more difficult it is to extract. Gold and silver are easy to extract (once found!) by washing (as we have seen in countless Westerns) whereas sodium, magnesium and potassium, which do not normally occur alone naturally but in compounds, present most difficulty.

(3) The higher the position in the series of a metal the greater is its reaction with air and water. Potassium and sodium are so reactive that they are normally kept under oil to prevent them coming into contact with the atmosphere until required. Iron rusts rapidly when in contact with damp air but gold and silver remain untarnished by either air or water, unless the atmosphere contains an unusually large proportion of sulphur compounds as in an urban environment. Magnesium, aluminium and zinc react with air to form oxides, but once formed they protect the metal to some extent and so these materials can be used safely for constructional purposes. Zinc although higher in the series is often used to protect iron, because zinc oxide tends to protect the zinc and therefore the iron from further corrosion, whereas hydrated ferric oxide (rust) which forms on iron does not protect it.

(4) When two metals are placed in an electrolyte a simple cell is set up and the one higher in the series becomes the negative electrode and is gradually dissolved. Thus in the example of the simple cell given earlier, zinc, which is higher than copper, becomes negative and also dissolves. Simple cells are often set up under natural conditions and lead to corrosion, which can be dangerous. In riveting steel plates together, care must be taken that the rivets are not made of a metal lower in the series than iron for, if they are, under damp atmospheric conditions or particularly in water (the hull of a ship, for example) the plates begin to dissolve in the region

of the rivets as simple cell action commences. Similarly, if metals are coated with others for protection, the coating metal must be higher in the series. Zinc-coated iron (galvanised iron) is acceptable, because if the coating is scratched and the iron exposed, the simple cell action between zinc–iron and electrolyte (atmosphere or water) will cause the zinc to dissolve before the iron. Tin plating of iron would not be suitable under such conditions. Corrosion of underground pipelines may be prevented by burying at intervals and in close proximity pieces of a metal higher in the series (for example, magnesium if the pipelines are iron). The higher metal then dissolves before the pipe material if the soil water content acts as an electrolyte, causing simple cell action.

ACIDS AND ALKALIS

An acid was originally defined as a compound containing hydrogen, some or all of which could be replaced by a metal to form a new compound known as a salt. Thus hydrochloric acid containing hydrogen and chlorine can be converted to sodium chloride (common salt) containing sodium and chlorine. Sulphuric acid, which contains sulphur, hydrogen and oxygen, can be changed to either sodium sulphate or sodium bisulphate, both of which are salts. Sodium sulphate contains sodium, sulphur and oxygen and sodium bisulphate contains these elements together with some hydrogen not replaced by sodium. There are, of course, many other acids but the same property is held by all of them.

In solution an acid ionises, producing positive ions containing hydrogen and oxygen, called *hydroxonium* ions. A modern definition of an acid is a compound which, in solution with water, produces hydroxonium ions as the *only* positive ions.

Acids have many properties in common. Some of these are listed below.

(1) Almost all acids react with carbonates and bicarbonates. A carbonate is a compound containing carbon and oxygen, the best-known example probably being sodium carbonate or 'washing soda'.

(2) Acid solutions turn *blue litmus* red, *methyl orange* pink and render *phenolphthalein* colourless. (These substances—blue litmus, methyl orange and phenolphthalein—used to detect acids are called *indicators*.)

(3) Dilute acids have a sour taste. The best-known acids here include citric acid (lemons), lactic acid (sour milk) and acetic acid (vinegar). It is obviously not advisable to test this property with more corrosive acids such as hydrochloric and sulphuric!

(4) Most acids react with the higher elements in the electrochemical series (magnesium, iron, etc.) and hydrogen is released. As might be expected reaction with sodium and potassium is violent and can be dangerous if not carefully controlled.

Compounds that react with acids to produce salt and water only are called *bases*. A base *neutralises* an acid removing its acidity, that is, changing its properties characteristic of an acid. A soluble base (one which can be dissolved in water to produce a solution) is called an *alkali*. Common alkali compounds are the hydroxides of sodium, barium, calcium, potassium and ammonia. A hydroxide molecule contains the metallic element and also hydrogen and oxygen. Ammonia, which is a compound of nitrogen and hydrogen, is not of course an element nor is it metallic, but because it establishes positive ions in solution, it is classified with the metals. Alkalis have the following properties.

(1) Alkalis neutralise acids to release a salt and water only.

(2) Alkalis have little effect on metals unless they are particularly strong; for example, a strong alkali will react with aluminium and release hydrogen. (Note that by the former definition this would have placed strong alkalis in the acid category. This is one reason for the more modern definition.)

(3) Alkalis also affect the indicators described earlier: they turn red litmus blue, phenolphthalein pink and methyl orange remains yellow.

(4) Alkalis are generally slippy if handled. Strong alkalis are good solvents for certain oils and grease.

CHEMICAL EQUATIONS

All the chemical reactions so far discussed have been described using words only. Normally a form of 'chemical shorthand' is used, using symbols and describing the reaction symbolically in the form of an equation. Symbols of the more commonly used elements are as follows.

Aluminium	Al	Mercury	Hg
Antimony	Sb	Molybdenum	Mo
Argon	A	Neon	Ne
Arsenic	As	Nickel	Ni
Beryllium	Be	Nitrogen	N
Bismuth	Bi	Oxygen	O
Boron	B	Phosphorus	P
Bromine	Br	Platinum	Pt
Cadmium	Cd	Potassium	K
Caesium	Cs	Radium	Ra
Calcium	Ca	Selenium	Se
Carbon	C	Silicon	Si
Chlorine	Cl	Silver	Ag
Chromium	Cr	Sodium	Na
Cobalt	Co	Strontium	Sr
Copper	Cu	Sulphur	S
Gold	Au	Thorium	Th
Helium	He	Tin	Sn
Hydrogen	H	Tungsten	W
Indium	In	Uranium	U
Iodine	I	Vanadium	V
Iron	Fe	Xenon	Xe
Lead	Pb	Yttrium	Y
Magnesium	Mg	Zinc	Zn
Mangenese	Mn		

(Some of the symbols are derived from Latin names, which explains Fe for iron, Sn for tin, etc.)

In describing compound molecules a particular notation is used to show atom content. For example, sulphuric acid, which has the formula H_2SO_4, contains per molecule two hydrogen atoms, one sulphur atom and four oxygen atoms. Similarly, hydrochloric acid, HCl, contains one hydrogen atom and one chlorine atom per acid molecule.

In using a chemical equation care must be taken to ensure that a balance exists between left- and right-hand sides and that the numbers of a particular atom are equal on both sides. The following examples should be carefully studied.

$$Mg + H_2SO_4 = MgSO_4 + H_2$$

Magnesium and sulphuric acid gives magnesium sulphate and hydrogen.

$$CuO + H_2SO_4 = CuSO_4 + H_2O$$

Copper oxide and sulphuric acid gives copper sulphate and water. (This is an example of acid neutralisation by a base to give a salt and water.)

$$2Al + 2NaOH + 2H_2O = 2NaAlO_2 + 3H_2$$

Aluminium and sodium hydroxide and water gives sodium aluminate and hydrogen. (An example of the action of an alkali solution on aluminium.) Note here that there is an adjustment of numbers preceding the collection of symbols denoting a molecule so that the equation 'balances'. In more detail this equation tells us that two molecules of aluminium with two molecules of sodium hydroxide and two molecules of water give two molecules of sodium aluminate and *three* molecules of hydrogen. (Remember that hydrogen atoms go around in pairs! A hydrogen molecule contains two atoms and is denoted by H_2 using symbols.)

The formation of rust is described by the following equations

$$Fe + H_2O + CO_2 = FeCO_3 + H_2$$

Iron and water and carbon dioxide gives ferrous carbonate and hydrogen.

$$4FeCO_3 + O_2 + 2CO_2 = 2Fe_2(CO_3)_3$$

Ferrous carbonate and oxygen and carbon dioxide gives ferric carbonate.

$$2Fe_2(CO_3)_3 + 3H_2O = 2Fe_2O_3 \cdot 3H_2O + 6CO_2$$

Ferric carbonate and water gives hydrated ferric oxide (rust) and carbon dioxide. Note the complex molecule $Fe_2(CO_3)_3$, which contains two atoms of iron and three sets of atoms, each set containing one carbon atom and three oxygen atoms. Such 'sets' are called *radicals*. Other commonly occurring radicals include sulphates (SO_4), nitrates (NO_3) and the ammonium radical (NH_4).

Chemical equations, as can be seen, tell us a great deal about chemical reactions (but not their circumstances, that is, whether or not heat is required, for example) and show how the various elements are rearranged to form new compounds during such reactions. They can also be used to indicate what weights of substances are involved in reactions; this technique is outside the scope of this book, since knowledge of it is not required for the attainment of the objectives listed in this Unit.

ASSESSMENT EXERCISES

Long Answer

11.1 (a) Explain the difference between a chemical change and a physical change and give *one* example of each. (b) Distinguish between the characteristics of elements, compounds and mixtures and give *one* example of each.

11.2 Explain what is meant by the following terms used in connection with basic atomic theory: (a) electron, (b) atom, (c) molecule, (d) atomic number, (e) ion.

11.3 Explain the meaning of the following terms: (a) solution, (b) suspension, (c) soluble, (d) solute, (e) solvent, (f) saturated solution, (g) solubility, (h) alloy. Give one example of an alloy.

11.4 (a) With the aid of simple diagrams explain what is meant by a crystal and give *three* examples of materials with a crystalline structure. (b) Give one example of an amorphous substance.

11.5 (a) Give the main characteristics of the element oxygen and state *five* of its uses. (b) Briefly describe what is meant by oxidation and, without the use of chemical formulae, the process of rusting.

11.6 (a) Explain the term electrolysis. (b) Give *three* examples of electrolysis, stating the materials used as anode, cathode and electrolyte and briefly describing the reactions at each electrode. (c) Describe *one* use of the process of electrolysis.

11.7 (a) Explain what is meant by the 'electrochemical series'. (b) Distinguish between the characteristics of alkali metals, alkaline earths, heavy metals and noble metals.

11.8 Distinguish between the characteristics of acids and alkalis. A detailed list is required.

11.9 Write down the chemical equations for the reactions listed below and explain what is meant by any *one* of the equations in terms of atoms and molecules: (a) magnesium added to sulphuric acid, (b) copper oxide and sulphuric acid, (c) aluminium, sodium hydroxide and water.

Short Answer

11.10 Give an example of a chemical change.

11.11 Give one example of a physical change.

11.12 Name four elements.

11.13 Define the term 'atom'.

11.14 Define the term 'molecule'.

11.15 Give one example of a compound.

11.16 Give one example of a mixture.

11.17 Define 'atomic number'.

11.18 Define the term 'solution'.

11.19 What is meant by the word 'alloy'?

11.20 Define the term 'amorphous'.

11.21 Give three characteristics of oxygen.

11.22 Briefly explain the term 'electrolysis'.

11.23 What is the reaction at the anode when two carbon rods are placed in hydrochloric acid and electrolysis occurs?

11.24 Name the main components of a simple cell.

11.25 Are alkali metals higher or lower than noble metals in the electrochemical series?

11.26 List the effects of acids on three separate indicators.

11.27 List the effects of alkalis on three separate indicators.

11.28 Give the chemical symbols for cadmium, copper, gold, magnesium, manganese, phosphorus, potassium and uranium.

11.29 Give the chemical symbols for hydrochloric acid, sulphuric acid, copper oxide and sodium hydroxide.

Multiple Choice

11.30 When hydrogen and oxygen combine to form water the reaction is
A. a physical change   B. a chemical change   C. a mixture of both chemical and physical change   D. neither a chemical nor a physical change

11.31 The following list contains three compounds and one mixture. Select the mixture
A. salt and water   B. sodium and chlorine   C. silicon and oxygen   D. hydrogen and oxygen

11.32 The atomic number of an element is the number of
A. electrons in an atom of the element   B. particles in an atom of the element   C. atoms making up a molecule of the element   D. orbits in an atom of the element

11.33 When sugar is completely dissolved in water the resulting clear liquid is called a
A. suspension   B. solute   C. solvent   D. solution

11.34 Select the amorphous material in the following list
A. diamond   B. graphite   C. glass   D. fluorite

11.35 If two carbon rods are placed in a salt solution and electrolysis takes place the substance released at the anode is
A. salt   B. sodium   C. chlorine   D. hydrogen

11.36 In the electrochemical series
A. gold is higher than silver   B. platinum is higher than mercury   C. heavy metals are below noble earths   D. alkali metals are above alkaline earths

11.37 Acid solutions
A. turn red litmus blue   B. render methyl orange yellow   C. have a sour taste   D. do not contain hydroxonium ions

11.38 Alkalis
A. react violently with metals   B. turn red litmus blue   C. render methyl orange pink   D. make phenolphthalein colourless

11.39 The chemical formula for sulphuric acid is
A. HSO_4 B. H_2SO_3 C. H_2SO_2 D. H_2SO_4

12 Cells and Batteries

OBJECTIVES

All the objectives should be understood to be prefixed by the words 'The expected learning outcome is that the student . . .'

I23 Knows the concepts of e.m.f. and internal resistance.
- 23.1 Describes the potential difference (voltage) of a source on no-load as the e.m.f.
- 23.2 Uses a high-resistance voltmeter to measure the e.m.f. of a dry cell and a battery of dry cells.
- 23.3 Explains the reasons for the voltmeter in 23.2 having a high resistance.
- 23.4 Defines internal resistance.
- 23.5 Measures the effect of load current on terminal p.d. and hence determines internal resistance.

I24 Describes secondary cells.
- 24.1 Explains the difference between primary and secondary cells.
- 24.2 Describes the charging and discharging of a simple lead–acid cell.
- 24.3 Measures, using a high-resistance voltmeter, the e.m.f. of charged lead–acid cells, (a) singly (b) in series (c) in parallel.
- 24.4 Measures the effect of load current on terminal p.d. and hence determines internal resistance of the examples in 24.3.
- 24.5 Labels, given diagrams, the main parts of (a) lead–acid cells (b) alkaline cells (c) mercury cells.

Batteries have become an essential feature of modern daily life. The 'throw-away' variety, correctly called *primary* batteries, are used in radios, torches, clocks, gas-lighters, watches, calculators, toys, tape recorders, doorbells, hearing aids, to name but a few applications. Rechargeable batteries, correctly called *secondary* batteries, may also find their way into some of these items but, in the form of the lead–acid variety particularly, they are mainly used in cars and other vehicles for starting purposes, in electric vans and buses for traction and as standby emergency supplies in the event of mains electricity failure (see figure 12.1, courtesy of Chloride Industrial Batteries Ltd).

A battery of either type consists of a number of cells connected in series (or occasionally in series–parallel) each cell containing two electrodes—one positive and one negative—an electrolyte and other chemicals as required. The basic theory of a simple cell consisting of a copper (positive) electrode and a zinc (negative) electrode immersed in an electrolyte of sulphuric acid, was given in chapter 11. As stated there the simple cell is not of practical use because of the formation of an insulating layer of hydrogen at the copper electrode. Additional chemicals to act as a *depolariser* must be added to make the cell useful.

PRIMARY CELLS

A primary battery, consisting of interconnected primary cells, is thrown away when discharged; it cannot be recharged to any useful extent. (Battery manufacturers warn that attempts to recharge primary batteries can lead to a build up of dangerous gas pressures inside them.) Three of the most commonly used primary cells will be described: the Leclanche cell, the mercury cell and the alkaline manganese cell.

Figure 12.1

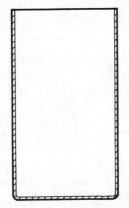
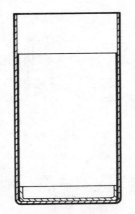
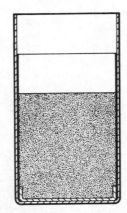
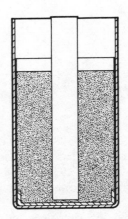
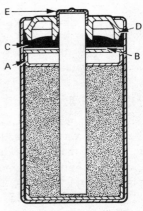

(a) In the round cell the container is a zinc cup, which, besides containing the constituents of the cell, also acts as the anode

(b) An absorbent paper separator is inserted before the cathode to prevent the anode and cathode touching

(c) The cathode is powdered manganese dioxide, mixed with carbon black to make it conducting, and wetted with electrolyte solution

(d) A carbon rod is used to make electrical contact between the cathode and the positive terminal of the cell

(e) A cardboard tamping washer (A) is placed over the cathode mix and then a top collar (B). A bitumen seal (C) on the top collar prevents evaporation of water from the electrolyte. The cell is closed at the top with a plastic top cover (D) and a metal top cap (E) and then fitted with a metal jacket (not shown)

Figure 12.2   The Construction of the Leclanché Cell (Courtesy of the Ever Ready Co.)

The Leclanché Cell

The original Leclanché cell was invented by Georges Leclanché, a French engineer, in 1866. It was immediately successful and by 1868 over twenty thousand were in use in telegraph systems. It was assembled in a glass jar and consisted of a positive electrode of carbon surrounded by manganese dioxide to act as a depolariser and, contained in a porous pot within the outer glass container, a negative zinc electrode and an electrolyte of ammonium chloride solution thickened by the addition of sand or sawdust. Although many variants have been successfully introduced, including using coke, graphite and carbon black in place of carbon for the cathode, and electrolytically prepared materials in place of the manganese-based depolariser, the basic materials of the Leclanché cell remain the same. Figures 12.2, 12.3 and 12.4 show the construction of the modern version of the Leclanché cell. Its performance will be discussed later.

The Mercury Cell

The use of mercuric oxide and zinc in a potassium hydroxide electrolyte was proposed as early as 1886 but the construction methods of the time produced a cell with a high-resistance positive electrode. Work conducted in the 1930s by Samuel Ruben produced the basis for the modern cell, one form of which is illustrated in figure 12.5.

Mercury cells are produced in cylindrical and flat pellet or button construction. The chemical content is the same in each case. A typical flat pellet construction is shown in figure 12.5. The anode is formed from high-purity amalgamated zinc powder pressed into pellets. The cathode is a mixture of mercuric oxide and graphite (for conductivity) and is separated from the anode by an ion-permeable barrier. The electrolyte is a strongly alkaline

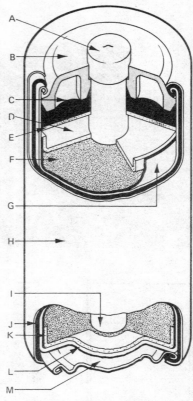

D. Top Washer. This functions as a spacer and is situated between the depolariser mix and the top collar.
E. Top Collar. This centralises the carbon rod and supports the bitumen sub-seal.
F. Depolariser. This is made from thoroughly mixed, high quality materials. It contains manganese dioxide to act as the electrode material and carbon black for conductivity. Ammonium chloride and zinc chloride are other necessary ingredients.
G. Paper Lining. This is an absorbent paper, impregnated with electrolyte, which acts as a separator.
H. Metal Jacket. This is crimped on to the outside of the cell and carries the printed design. This jacket resists bulging, breakage and leakage and holds all components firmly together.
I. Carbon Rod. The positive pole is a rod made of highly conductive carbon. It functions as a current collector and remains unaltered by the reactions occurring within the cell.
J. Paper Tube. This is made from three layers of paper bonded together by a waterproof adhesive.
K. Bottom Washer. This separates the depolariser from the zinc cup.
L. Zinc Cup. This consists of zinc metal, extruded to form a seamless cup. It holds all the other constituents making the article clean, compact and easily portable. The cup is also the anode and when the cell is discharged part of the cup is consumed to produce electrical energy.
M. Metal Bottom Cover. This is made of tin plate and is in contact with the bottom of the zinc cup. This gives an improved negative contact in the torch or other equipment, and seals off the cell to increase its leak resistance.

A. Metal Top Cap. This is provided with a pointed pip to secure the best possible electrical contact between cells.
B. Plastic Top Cover. This closes the cell and centralises the positive terminal.
C. Soft Bitumen Sub-Seal. A soft bitumen compound is applied to seal the cell.

The Round Cell Battery (Courtesy of the Ever Ready Co.)

Figure 12.3

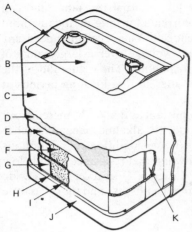

A. Protector Card. This protects the terminals and is torn away before use.

B. Top Plate. This plastic plate carries the snap fastener connectors and closes the top of the battery.

C. Metal Jacket. This is crimped on to the outside of the battery and carries the printed design. This jacket helps to resist bulging, breakage and leakage and holds all components firmly together.

D. Wax Coating. This seals any capillary passages between cells and the atmosphere, so preventing the loss of moisture.

E. Plastic Cell Container. This plastic band holds together all the components of a single cell.

F. Depolariser. This is a flat cake containing a mixture of manganese dioxide as the electrode material and carbon black for conductivity. Ammonium chloride and zinc chloride are other necessary ingredients.

G. Paper Tray. This acts as a separator between the mix cake and the zinc electrode.

H. Carbon Coated Zinc Electrode. Known as a Duplex Electrode, this is a zinc plate to which is adhered a thin layer of highly conductive carbon which is impervious to electrolyte.

I. Electrolyte Impregnated Paper. This contains the electrolyte and is an additional separator between the mix cake and the zinc.

J. Bottom Plate. This plastic plate closes the bottom of the battery.

K. Conducting Strip. This makes contact with the negative zinc plate at the base of the stack and is connected to the negative socket at the other end.

The Flat Cell Battery (Courtesy of the Ever Ready Co.)

Figure 12.4

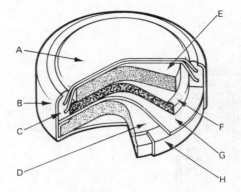

A. double top
B. cell can
C. plastic sealing grommet
D. depolariser pellet
E. zinc anode pellet
F. electrolyte in absorbent
G. synthetic separator
H. depolariser sleeve

The Mercury Cell (Courtesy of the Ever Ready Co.)

Figure 12.5

aqueous solution, usually potassium hydroxide, with a high ionic conductivity. It provides a source of water and hydroxyl ions*, both of which are essential for the electrochemical reactions occuring in the cell during discharge. The inner cell top is plated to provide an internal surface electrochemically compatible with the zinc anode (to minimise corrosion and harmful evolution of

* A hydroxyl ion is negative and has the formula 2OH.

hydrogen). The cell containers are made of nickel-plated steel, since this material is not easily corroded by alkaline electrolytes.

The performance of the mercury cell will be discussed later.

The Alkaline Manganese Cell

Alkaline manganese cells are similar in construction to mercury cells and again are available in cylindrical and button form with cell containers of electroplated steel. The cylindrical form is illustrated in figure 12.6.

The positive terminal is formed by a stud at the top of the cell in the cylindrical construction and is in contact with the depolariser via the steel case. The depolariser or cathode is a high-density mixture of manganese dioxide and graphite compressed into cylinders which fit around the anode, but separated from it by an electrolyte-absorbent material.

The anode consists of powdered zinc, with a large surface area and formed into a paste with the potassium hydroxide electrolyte. The anode is connected to the negative terminal at the base of the cell by an internal contact. The performance of the cell is discussed below.

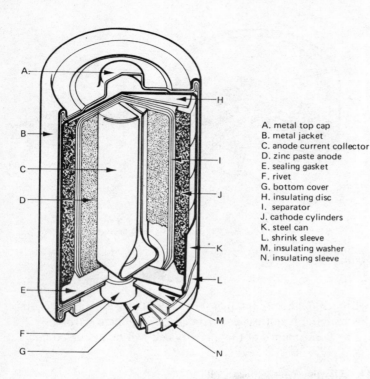

A. metal top cap
B. metal jacket
C. anode current collector
D. zinc paste anode
E. sealing gasket
F. rivet
G. bottom cover
H. insulating disc
I. separator
J. cathode cylinders
K. steel can
L. shrink sleeve
M. insulating washer
N. insulating sleeve

The Alkaline Manganese Cell (Courtesy of the Ever Ready Co.)

Figure 12.6

Performance of Primary Dry Cells

Leclanché dry cells are available in 'power pack' form for use in transistorised equipment where a relatively low current is required intermittently and in 'high power' form in equipment which takes comparatively heavy current, for example, recorders, shavers and cine cameras. The single cell has an open-circuit (unloaded) e.m.f. of 1.6 V falling to between 1.2 V and 1.4 V when fully discharged. The internal resistance of the cell across which voltage is lost when the cell is loaded is relatively high compared to the other primary cells described. Internal resistance and its measurement is considered in more detail at the end of the chapter.

The mercury cell has a very low internal resistance and can provide steady voltage at high currents for long periods without the 'rest' required by the Leclanche cell. When a cell or, indeed, any power supply can provide steady voltage over a large range of current it is said to have good *regulation*. The regulation of the mercury cell remains good even after storing for long periods at temperatures of 20 °C and above.

Although its voltage is not as constant as that of the mercury cell (that is, its regulation is not as good), the alkaline manganese cell is able to give a useful life under extreme conditions of continuous discharge. This cell has a very low internal resistance (well below 1 ohm) and will provide heavy currents. Operation over a wide temperature range (− 20 °C to 70 °C) is possible and the cell is little affected by long storage periods.

SECONDARY CELLS

Secondary batteries consist of a number of interconnected secondary cells. They can be recharged once their output voltage falls below a useful level and the charge–discharge process can be repeated many hundreds of times. They are available either in 'wet' or 'dry' form, the electrolyte being either a liquid or paste respectively; in the 'dry' form the electrolyte may also be a jelly. Dry cells are sealed and may be used in any position, wet cells must be used in an upright position and capped openings are made available for 'topping up' the liquid electrolyte.

The Lead–Acid Cell

The modern lead–acid cell has its roots in the experimental work of scientists such as Ritter, Grove, Faraday and Planté in the early and mid-nineteenth century. The discovery in 1881 of the process of forming active materials from lead oxides, coupled with the emergence of the electric dynamo as a quick means of recharging, proved a great stimulus to the commercial production and extended use of storage batteries.

The lead–acid cell has positive and negative plates in which the active materials are lead peroxide and spongy lead respectively, immersed in an electrolyte of dilute sulphuric acid. There are two types of plate: the Planté and the Faure, or pasted, type. The Planté plate consists of a sheet of lead on which the active material is formed electrochemically from the lead of the plate itself. There are various kinds of pasted plate, the active material being in the form of lead oxides applied during manufacture. The 'flat' pasted plate consists of cast lead alloy grids with a lattice of intersecting ribs into which the active material is pasted. This paste consists of a carefully chosen mixture of lead oxide and dilute sulphuric acid, which reacts with the oxide powders to set hard within the meshwork of the grid structure. The grids surrounding the active

material act as conductors for the electric current. Another design of plate, called 'tubular positive', is made up of rods of antimonial lead surrounded by tubes of an inert porous material such as hard rubber or woven Terylene. The spaces between the rods and tubes are filled with lead oxide powder. Immersion of the plate in dilute

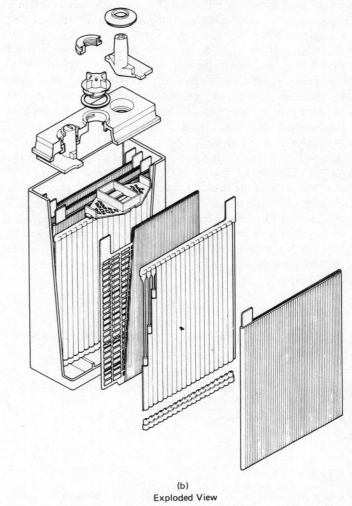

filling plug

post rings

cell lid

sealing gaskets

combined separator guard and acid level indicator

positive plates

negative plates

porvic sleeves (microporous PVC)

cell container

(a)
Lead-Acid Secondary Cell (Courtesy of Chloride Industrial Batteries Ltd)

(b)
Exploded View

Figure 12.7

sulphuric acid produces an electrochemical reaction which forms a hard 'pencil' of active material.

The cell is constructed by connecting together plates of similar polarity and interleaving them. Plates of opposite polarity are kept apart by separators made from a material which is porous but resistant to acid and chemical action. Typical materials are certain woods, glass wool, perforated or porous plastic or rubber. See figure 12.7.

In a fully charged cell the positive active material is lead peroxide, PbO_2, and the negative active material is spongy lead, Pb. The electrolyte is dilute sulphuric acid, H_2SO_4. On discharge the lead peroxide gives up oxygen and combines with the sulphate radical of the acid to form lead sulphate, $PbSO_4$. The spongy lead also combines similarly with the acid to give lead sulphate on the negative plates as well. The oxygen liberated from the lead peroxide combines with hydrogen in the acid to form water, thus diluting the acid further. The specific gravity of the electrolyte (the ratio of the weight of a given volume of electrolyte divided by the weight of an equal volume of water at the same temperature) falls and the cell voltage also falls. On recharge, carried out by connecting a suitable d.c. supply to the battery (supply positive to battery positive, supply negative to battery negative) the active materials are reformed from the lead sulphate. The process is summarised in the following chemical equation

$$PbO_2 + 2H_2SO_4 + Pb \underset{\xleftarrow{\text{charge}}}{\xrightarrow{\text{discharge}}} 2PbSO_4 + 2H_2O$$

Lead peroxide and acid and lead gives lead sulphate and water.

Operation and maintenance of lead–acid batteries will be described later.

The Alkaline Cell

The basic development of alkaline cells dates from 1881 when Lalande and Chaperon patented a cell using a copper oxide positive electrode with a zinc negative electrode in potassium or sodium hydroxide electrolyte. In 1889 Thomas Edison patented a modified version of the Lalande–Chaperon cell and from around 1900 experimented with nickel oxide and iron as electrode materials. Later developments resulting from work carried out by Edison and Jungner, who worked independently, laid the basis of modern nickel–cadmium cells now available in sealed form.

Over the years a number of different types of alkaline secondary cells have become available, but their basic content is similar. The electrolyte is invariably potassium hydroxide, the positive electrode contains nickel or its compounds and the negative electrode contains iron or cadmium, or both, together with their compounds.

One form of 'wet' cell widely used in the British Armed Services in recent years has plates made up as follows. The active material in the positive plate is nickel hydroxide mixed with graphite (to improve conductivity) while that in the negative plate is a mixture of the oxides of cadmium and iron. The active materials are compressed into briquettes under high pressure and held in steel 'pockets' in nickel-plated steel frames. A number of cells is interconnected to form a battery in much the same way as in the lead–acid battery, with ebonite being used as separators. The container is made from steel, plated to prevent rust, and the electrolyte is potassium hydroxide dissolved in distilled water. Cells of this type, particularly where iron or its compounds are used in the negative plate active material, are often referred to as NIFE cells—after the chemical symbols for nickel–iron—or Edison cells, after the man who did so much work to develop the basic idea.

A typical nickel–cadmium cell is shown in figure 12.8. The contruction shown is of a cylindrical cell but button cells are also available; electrochemically the forms are identical. As can be seen the cell is of the 'dry' variety and is sealed. Basically the nickel–cadmium cell is similar to the nickel–iron cell but with cadmium being used as the negative electrode. When charged the positive electrode is nickelic hydroxide changing to nickel oxide on discharge, and the negative electrode is cadmium when the cell is charged changing to cadmium hydroxide on discharge. The electrolyte, potassium hydroxide, remains unchanged during charge or discharge. The electrodes consist of a perforated nickel or

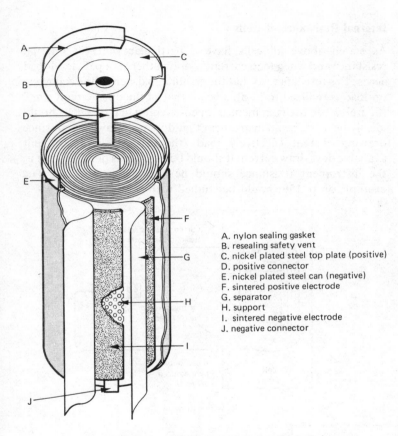

A. nylon sealing gasket
B. resealing safety vent
C. nickel plated steel top plate (positive)
D. positive connector
E. nickel plated steel can (negative)
F. sintered positive electrode
G. separator
H. support
I. sintered negative electrode
J. negative connector

The Nickel-Cadmium Cell (Courtesy of the Ever Ready Co.)

Figure 12.8

nickel-plated mesh covered by a microporous pure nickel matrix (achieved by a process called 'sintering'). The spaces in the mesh are filled with the electrode active materials following high vacuum treatment to ensure that all the pores are completely filled. Nickel strips are welded to the electrodes and make contact with the plated steel battery casing and top. The electrode separators are formed from polyamide material specially selected for long-term physical and chemical stability. The cell is hermetically sealed, once

the electrolyte has been added, by means of a creep-resistant insulating nylon gasket pressed between the top of the can and the top plate.

Operation and Maintenance of Secondary Cells and Batteries

Before describing operation and maintenance of secondary cells and batteries the following definitions are needed.

Open-circuit E.M.F. and Terminal P.D.

As with all voltage sources the voltage off-load is higher than the voltage on-load because of the voltage drop across the internal resistance of the cell; this is further discussed at the end of the chapter. The cell or battery voltage off-load is called the open-circuit e.m.f.; on-load it is called the terminal p.d. Cell or battery voltage on- and off-load depends on the state of charge, cell temperature and the age of the cell. In addition, the terminal p.d. depends on the size of the charge or discharge current.

Capacity

The capacity of a cell or battery is the amount of charge available during discharging from the fully charged to the fully discharged condition. The coulomb is too small a unit for practical use in this case; the *ampere-hour* is used instead. The ampere-hour (symbol A h) is the charge moved by a current of one ampere flowing for one hour (as opposed to the coulomb which is the charge moved by a current of one ampere flowing for one second.) Capacity is the product of discharge current and time of discharge in hours, but because capacity depends on rate of discharge—the higher the rate the lower the capacity—a charging rate is also given when the capacity is given by the manufacturer. Thus a capacity of 500 A h at the 10 hour rate means that 50 A can be supplied continuously for 10 hours. It does not mean that 100 A can be supplied for 5 hours, however, since the rate of discharge current would then be increased. It is usually a reasonable assumption that the same capacity applies at lower discharge rates, so that the 500 A h, 10 h rate quoted would probably mean we could obtain

25 A for 20 h or 10 A for 50 h. Normally a secondary cell or battery is charged at the rate quoted, that is, a 10 h rate battery is charged over 10 h, but higher rates of charge can be tolerated in many instances especially in the case of nickel–cadmium cells.

Specific Gravity: the Hydrometer

The specific gravity of a liquid was defined on p. 152. It is measured using an instrument called a *hydrometer*, which consists of a graduated float contained in a glass syringe into which the liquid is drawn by means of a rubber bulb (figure 12.9). Sometimes the decimal point is missed off the graduated float and the figures must then be divided by 1000 to give the specific gravity.

Conditions of charge and discharge, specific gravity and terminal voltage levels for alkaline and lead–acid cells are summarised in table 12.1 together with their respective advantages and disadvantages and a list of precautions to be observed.

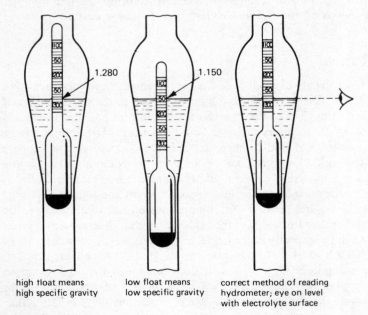

high float means high specific gravity

low float means low specific gravity

correct method of reading hydrometer; eye on level with electrolyte surface

Reading a Hydrometer (Courtesy of Chloride Industrial Batteries Ltd)

Figure 12.9

Internal Resistance of Cells

As stated above all cells have a particular value of internal resistance and when load current is drawn there is a p.d. developed across this resistance, so that the terminal p.d. across the cell when on-load is reduced to a value *below* that of the open-circuit e.m.f. (figure 12.10). Measurement of open-circuit e.m.f. of a cell should ideally be made by an instrument that does not draw current since drawing current effectively loads the cell. If the instrument available does draw current it should be as small as possible, that is, the instrument resistance should be as high as possible. The example on p. 156 should be studied with care.

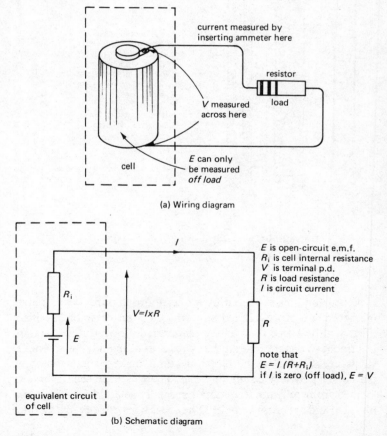

current measured by inserting ammeter here

resistor

load

V measured across here

cell

E can only be measured *off load*

(a) Wiring diagram

R_i

E

I

$V = I \times R$

R

E is open-circuit e.m.f.
R_i is cell internal resistance
V is terminal p.d.
R is load resistance
I is circuit current

note that
$E = I (R + R_i)$
if I is zero (off load), $E = V$

equivalent circuit of cell

(b) Schematic diagram

Figure 12.10

Table 12.1 Notes on Operation and Maintenance

Cell	State	Cell Voltage	Specific Gravity	Remarks
Lead – Acid (wet cell)	Discharged	Not less than 1.8 V	According to maker's instructions: commonly 1.18	Both plates whitish grey lead sulphate.
	Charged (on load)	Approximately 2.7 V	Commonly 1.27	Positive plate: brown lead peroxide
	On charge		Commonly 1.27	Negative plate: slate grey lead if left discharged or discharged below specification plates become sulphated and cell may become permanently damaged. When diluting acid use only pure water. Add acid to water, never water to acid. Adversely affected by high rates of charge or discharge
Alkaline (wet cell)	Discharged	1.1 V approx	Commonly 1.19	Common fault: loss of capacity caused by undercharging or causing cells to stand idle.
	Charged (on load)	Between 1.2 V and 1.3 V	No change	Can withstand high charging and discharging currents without damage.
	On charge	1.7 V approx	No change	Top up with distilled water. Do not allow specific gravity to fall below specification by more than 0.03
Nickel – Cadmium ('dry' cell)	Charged	1.2 to 1.3 V		Maintenance-free. Do not discharge below 1 V per cell.
	Discharged	1.0 V		Follow manufacturer's instructions on recharging.
	On charge	1.45 V		May be used in any position. Can withstand large rates of charge and discharge over long periods
Lead – Acid ('dry' cell)				Maintenance-free

Example 12.1

A battery with an open-circuit e.m.f. of 6.2 V is connected to a resistive load of 100 Ω. The terminal p.d. is measured and is found to be 6.19 V. Calculate the internal resistance of the battery.
Solution The total e.m.f. available (*E* in figure 12.10) is 6.2 V

$$\text{circuit current} = \frac{\text{terminal voltage}}{\text{load resistance}}$$

$$= \frac{6.19}{100} \text{ A}$$

therefore

$$\text{total resistance of circuit} = \frac{6.2}{6.19/100}$$

$$= 100.16 \, \Omega$$

$$\text{internal resistance} = \text{total resistance} - \text{load resistance}$$
$$= 100.16 - 100$$
$$= 0.16 \, \Omega$$

ASSESSMENT EXERCISES

Long Answer

12.1 (a) Explain briefly the difference between primary and secondary cells. (b) With the aid of diagrams describe *one* version of a commercially available primary cell.

12.2 Describe the Leclanché cell and its development from invention to the modern commercial version.

12.3 Compare the properties of mercury cells and alkaline manganese cells. Using a diagram describe a modern commercial version of *one* of these cells.

12.4 Describe the modern version of a lead–acid battery. List the advantages of such a battery and briefly outline the points to be borne in mind in its care and maintenance.

12.5 Describe with the aid of diagrams a nickel–cadmium cell. List the advantages of such a cell and indicate where it might be used to advantage.

12.6 (a) Define specific gravity. How is this quantity measured? (b) List the characteristics including values of specific gravity of a lead–acid battery (i) in a charged condition, (ii) in a discharged condition.

12.7 Tabulate values of cell voltage and specific gravity of the following cells in charged and discharged conditions: (a) lead–acid, (b) alkaline, (c) nickel–cadmium. In the table include brief comments on points to be borne in mind during the operation and maintenance of these batteries.

12.8 Figure 12.11 shows a cut-away drawing of a round cell battery. List its component parts as indicated by the letters. Of what early cell is this a development?

12.9 (a) Define 'internal resistance', 'e.m.f.' and 'terminal p.d.' of a cell. (b) A battery with an open-circuit e.m.f. of 3 V is connected to a resistive load of 10 Ω. The terminal p.d. is measured and found to be 2.95 V. Calculate the battery internal resistance.

Short Answer

12.10 State the main difference between primary and secondary cells.

12.11 List the main components of the Leclanche cell.

12.12 List the main components of the mercury cell.

12.13 List the main components of the alkaline–manganese cell.

12.14 Compare the internal resistance of the Leclanche, mercury and alkaline–manganese cells.

12.15 List the materials in a lead–acid cell when fully charged.

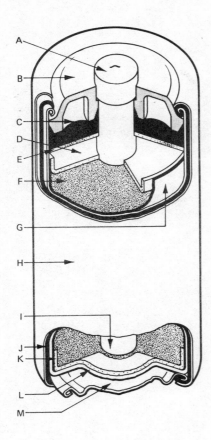

Figure 12.11

12.16 List the materials in a lead–acid cell when discharged.

12.17 Define capacity of a secondary cell.

12.18 Give the maximum and minimum values of specific gravity of a lead–acid cell when charged and discharged.

12.19 Define internal resistance of a cell.

Multiple Choice

12.20 The type of battery used in vehicles for starting purposes is usually
    A. lead–acid   B. Leclanché   C. Alkaline–manganese
D. mercury

12.21 In the mercury cell the anode is made of
    A. mercury   B. mercuric oxide   C. zinc   D. potassium hydroxide

12.22 The depolariser in an alkaline–manganese cell is mainly
    A. potassium hydroxide   B. manganese dioxide   C. powdered zinc   D. caustic potash

12.23 The internal resistance of the alkaline manganese cell lies between
    A. 0 and $1\,\Omega$ B. $1\,\Omega$ and $10\,\Omega$ C. $10\,\Omega$ and $100\,\Omega$ D. $100\,\Omega$ and $1000\,\Omega$

12.24 The unloaded e.m.f. of an unused Leclanché cell is
    A. $1.0\,V$ B. $1.2\,V$ C. $1.4\,V$ D. $1.6\,V$

12.25 In a fully charged lead–acid cell the positive active material is
    A. lead sulphate   B. spongy lead   C. lead peroxide   D. sulphuric acid

12.26 The negative active material in a discharged lead–acid cell is
    A. lead sulphate   B. spongy lead   C. sulphuric acid   D. lead peroxide

12.27 The negative electrode of an alkaline cell contains
    A. iron or cadmium   B. nickel   C. zinc   D. potassium hydroxide

12.28 A lead–acid battery is able to provide a current of 25 A for 10 h. The capacity at the 10-h rate for this battery (in A h) is
    A. 2.5   B. 250   C. 0.4   D. not determinable without further information

12.29 The colour of both plates in a discharged lead–acid cell is

A. black   B. whitish grey   C. brown   D. slate grey

12.30 The specific gravity of an alkaline cell when discharged is 1.19. When fully charged the specific gravity is

A. lower   B. higher   C. same   D. not determinable without further information

12.31 The internal resistance of a certain cell is 0.2. Its open-circuit e.m.f. is 1.5 V. When supplying a load current of 0.3 A its terminal p.d. will be

A. 1.5 V   B. 1.44 V   C. 1.56 V   D. not determinable without further information.

Answers to Numerical and Multiple Choice Assessment Exercises

CHAPTER 2

2.1 (a) 20.83; (b) 20; (c) 27.78; (d) 16.44 **2.2** (a) 5; (b) 0.14; (c) −0.27; (d) 5.42 **2.3** (a) 94 km/h, 80 km/h; (b) 86.67 km/h, 86.36 km/h, 83.3 km/h **2.4** 302.65 km/h, N 7°35′W **2.5** 3.2 m/s, 7.2 m/s **2.6** 77.46 s, 30.98 m/s **2.7** 51.6 m **2.8** 1.25 m/s², 7.25 m/s **2.9** 15.34 m/s **2.10** 22.62 kN **2.11** 192.36 kN, 555 m **2.12** 2:3, 5/7 h **2.13** (a) (i) 2.2 m/s² (ii) zero (iii) 1 m/s²; (b) 45 s (c) No. **2.14** 8 h 3 min **2.15** (a) 122.62 m (b) 49.05 m/s **2.16** (a) 49.32 km/h (b) 76 km/h **2.17** 6.22 m/s, 1.85 m/s², 1417.7 m **2.31** A **2.32** B **2.33** D **2.34** C **2.35** D **2.36** C **2.37** B **2.38** C **2.39** C **2.40** D

CHAPTER 3

3.1 1.85 N; 36.297 N **3.2** 37.74 N, 11°4′ **3.3** 905.9 N, 34°56′ **3.4** 214.3 N; 22′ **3.5** 6.99 N; 70°54′ **3.6** 1.17 m **3.7** 7 N, 9.67 N **3.8** 400 N **3.9** 1.35 m **3.10** 29.1 mm; 39.41 mm (from left-hand side and bottom side) **3.12** 46 m/s, 34° **3.13** 180 N, 70 N **3.14** 55.12 cm from side length 50 cm, 23.86 cm from side length 120 cm **3.15** 110.36 N; 134.89 N **3.16** 74.95 cm **3.17** 33.3 kPa, 2.62 × 10⁻² N **3.18** 116.78 kPa **3.19** 10 N **3.20** 4 m/s² **3.21** 1.25 t **3.22** 20 Nm **3.23** 0.25 m **3.24** 245 kN **3.25** 101.9 kg **3.26** 5 kPa **3.27** 1600 m² **3.28** 0.2 N **3.29** 66.2 kPa **3.30** 5 t/m³ **3.31** 5.89 N **3.32** 0.51 **3.33** 254.8 kg **3.34** D **3.35** B **3.36** D **3.37** A **3.38** B **3.39** C **3.40** D **3.41** C **3.42** A **3.43** C **3.44** B **3.45** A **3.46** A **3.47** B **3.48** D **3.49** D **3.50** D **3.51** D **3.52** B **3.53** C

CHAPTER 4

4.1 9.81 kN **4.2** 765 kg **4.3** 6.48 × 10⁷ N **4.4** 7 kg/m **4.5** 318.3 N **4.6** 18.4 kW **4.7** 36 kW **4.8** 18.87 s **4.9** 5.28 kW **4.10** 243.6 W **4.11** 5 kJ **4.12** 468.75 N **4.13** 12 m **4.14** 2.61 kW **4.15** 10 s **4.16** 1043.4 kJ **4.17** 94.8 % **4.18** 16.67 kW **4.19** 7.2 kW **4.20** 300 kJ **4.21** C **4.22** B **4.23** C **4.24** D **4.25** C **4.26** B **4.27** A **4.28** B **4.29** B **4.30** D

CHAPTER 5

5.1 4032 kJ **5.2** 30.8 kW **5.3** 0.23 **5.4** 14.34 kg, 3.4 kg **5.5** 0.376 **5.6** 24.13°C **5.7** 253 kg **5.8** 0.325, 0.003 25 kg **5.9** 4.46°C **5.10** 32°C **5.11** 0.27 kg **5.12** 21.6 × 10⁻⁶/°C **5.13** 2.04 mm **5.14** 1.6 × 10⁻⁸ **5.15** 22.02 m **5.16** 7.0067 × 10⁻² m² **5.17** 720.2 kJ **5.18** 0.14 kg **5.21** 168 kJ **5.22** 16.8 kJ/°C **5.23** 0.3 kg **5.24** 63 kJ **5.25** 3.35 MJ **5.26** 2257 kJ/kg **5.27** 0.000 18 m **5.29** D **5.30** A **5.31** D **5.32** C **5.33** B **5.34** D **5.35** C **5.36** A **5.37** D **5.38** C **5.39** D **5.40** A

CHAPTERS 6 AND 7

7.11 631.6 m **7.12** 1.2 MHz **7.19** B **7.20** A **7.21** C **7.22** D **7.23** B **7.24** C **7.25** C **7.26** A **7.27** B **7.28** C

CHAPTER 8

8.1 15.59 kN/m² **8.2** 344 kg **8.3** 56.5 kN/m² **8.4** 186.9 kN

8.5 $353\,\text{kN/mm}^2$ **8.6** 0.2% **8.7** 5.6×10^{-4}

8.8 $160\,\text{N/mm}^2$, $113.143\,\text{kN}$ **8.9** 7.27×10^{-5} **8.10** $70\,\text{mm}$

8.11 $10\,\text{kN/mm}^2$ **8.12** $6.9\,\text{mm}$; $400\,\text{kN/mm}^2$ **8.13** $1500\,\text{mm}^2$

8.16 $3.33\,\text{N/mm}^2$ **8.17** 0.0033 **8.19** 0.077 **8.20** $1\,\text{m}$

8.23 $300\,\text{kN/mm}^2$ **8.25** 0.15 **8.26** C **8.27** A **8.28** B **8.29** A

8.30 D **8.31** D **8.32** B **8.33** D **8.34** D **8.35** A

CHAPTER 9

9.1 $3.33\,\Omega$, $20\,\mu\text{A}$, $68\,\text{V}$, $7\,\text{A}$, $0.33\,\text{n}\Omega$, $13.6\,\text{V}$ **9.2** $55.5\,\Omega$

9.3 $83.33\,\text{mA}$, $17.86\,\text{mA}$, $10\,\text{mA}$ **9.4** $9.76\,\Omega$ **9.5** $14\,\text{V}$ **9.6** $100\,\Omega$

9.7 $2.389 \times 10^{-3}\,\text{S}$ **9.8** $0.11\,\mu\Omega$ **9.9** $650 \times 10^{12}\,\Omega$ **9.10** $111\,\Omega$

9.11 $127.67\,\Omega$ **9.12** $0.001\,75\,\Omega/\Omega^\circ\text{C}$ **9.13** (a) $0.163\,\text{W}$, $0.358\,\text{W}$, $0521\,\text{W}$; (b) $1.67\,\text{W}$, $0.758\,\text{W}$, $2.43\,\text{W}$ **9.14** $3.55\,\text{W}$, $5.14\,\text{W}$

9.15 $25.68\,\text{V}$, $0.5\,\text{W}$, $0.134\,\text{W}$ **9.16** $47.67\,\text{mA}$, $21.13\,\text{mA}$, $31.62\,\text{mA}$, $16.3\,\text{mA}$, $40.8\,\text{mA}$ **9.17** 30.64p **9.19** $8.33\,\text{mA}$

9.20 $103.4\,\text{V}$ **9.21** $2.62\,\text{k}\Omega$ **9.22** $9.2\,\text{k}\Omega$ **9.23** $2.17\,\text{k}\Omega$

9.24 $21.28\,\text{mW}$ **9.25** $0.68\,\text{W}$ **9.26** $10.8\,\text{W}$ **9.27** $6.66 \times 10^{-5}\,\text{S}$

9.29 $0.7242\,\Omega$ **9.30** $15\,\Omega$ **9.31** $4.23\,\text{kWh}$ **9.33** A **9.34** A

9.35 C **9.36** C **9.37** B **9.38** C **9.39** B **9.40** C **9.41** B

9.42 C **9.43** C **9.44** D

CHAPTER 10

10.3 (b) $4.5\,\text{V}$ **10.4** (b) $288\,\text{V}$ **10.5** (b) $53.33\,\text{H}$ **10.6** (b) $5.83\,\text{H}$

10.7 $0.8\,\text{V}$ **10.8** (b) $10.47\,\text{H}$ **10.9** (b) 0.67 **10.10** $82.5\,\text{V}$, $82.5\,\text{mH}$ **10.11** (a) $5\,\text{V}$; (b) $4\,\text{N}$; (c) $100\,\text{W}$ **10.12** $0.33\,\text{N}$

10.13 $0.04\,\text{T}$ **10.14** $2.25\,\text{V}$ **10.16** $20\,\text{ms}$ **10.17** $66.67\,\text{Hz}$

10.20 $10\,\text{H}$ **10.21** $1.05\,\text{V}$ **10.22** C **10.23** B **10.24** C **10.25** A

10.26 A **10.27** C **10.28** C **10.29** B **10.30** A **10.31** B

CHAPTER 11

11.30 B **11.31** A **11.32** A **11.33** D **11.34** C **11.35** C

11.36 D **11.37** C **11.38** B **11.39** D

CHAPTER 12

12.9 (b) $0.17\,\Omega$ **12.20** A **12.21** C **12.22** B **12.23** A **12.24** D

12.25 C **12.26** A **12.27** A **12.28** B **12.29** A **12.30** C

12.31 B